endless
UNIVERSE

Broadway Books | New York

endless
UNIVERSE

Beyond the Big Bang

Paul J. Steinhardt

and

Neil Turok

PUBLISHED BY BROADWAY BOOKS

Copyright © 2007 by Paul J. Steinhardt and Neil Turok

All Rights Reserved

A hardcover edition of this book was originally published in 2007
by Doubleday.

Published in the United States by Broadway Books, an imprint of
The Doubleday Publishing Group, a division of
Random House, Inc., New York.
www.broadwaybooks.com

Artwork created by Pier Gustafson

Book design by Donna Sinisgalli

Library of Congress Cataloging-in-Publication Data
Steinhardt, Paul J.
   Endless universe : Beyond the Big Bang / Paul J. Steinhardt
and Neil Turok.
      p. cm.
1. Cyclic universe theory (Cosmology) 2. Cosmology. I. Turok, Neil.
II. Title.
QB991.C92S83 2006
523.1'2—dc22

2006025256

ISBN 978-0-7679-1501-4

PRINTED IN THE UNITED STATES OF AMERICA

10 9 8 7 6 5 4

To Nancy and Corinne

Endless Universe was, in part, a family affair. The book could never have been completed without the love, patience, and support of Nancy and Corinne. Working on each side of the Atlantic, they were our most dedicated readers, our most exacting editors, our harshest critics, and our best morale boosters. Paul's sisters, Alice and Nancy Steinhardt, and brother-in-law Bob Sorensen, and Neil's parents, Ben and Mary, read early versions and gave vital guidance. Paul's sister-in-law Chana Shapiro meticulously reviewed the manuscript and helped us to resolve numerous structural issues, especially how to design a story told with two voices. Our children, Charlie, Joe, Will, and Cindy Steinhardt, and Ruby Turok-Squire, all helped to shape the text and the illustrations.

We also are extremely grateful to the numerous colleagues who took time to read all or parts of the manuscript and to give us their valuable advice, including Andreas Albrecht, Brian Greene, Stephen Hawking, Burt Ovrut, Roger Penrose, Martin Rees, Michael Turner, and Edward Witten. Daniel Baumann, Jean-Luc Lehners, and Paul McFadden carefully reviewed the physics presentation throughout the manuscript. Michael Olenick contributed many useful comments. Peter Galison, Gerald Holton, Peter Schaefer, and Jeffrey Tigay provided guidance on the historical issues discussed in Chapter 8. We thank Carl Feinberg for suggesting that we consider the relationship between the cyclic model and Asimov's short story *The Last Question*.

Many have contributed to our scientific and personal lives, as well as to the ideas described herein. The imprint of our mentors, Richard Feynman, Sidney Coleman, David Olive, and Tom Kibble, can be found throughout this book. For Paul, Alan Guth was profoundly important in drawing him into the field of cosmology (see Chapter 5). He, Michael Turner, and Andreas Albrecht were influencial early in Paul's career and have been longstanding friends ever since. Many at Cambridge and Princeton have provided invaluable guidance, including John and Neta Bahcall, Gary Gibbons, David Gross, Stephen Hawking, Jerry Ostriker, Lyman Page, Jim Peebles, Malcolm Perry, Martin Rees, Nathan Seiberg, David Spergel, Scott Tremaine, David Wilkinson, and Edward Witten.

In developing the cyclic model, we have been fortunate to have a cadre of outstanding collaborators. Burt Ovrut, a longtime collaborator and friend of Paul's, started us down the path toward the cyclic model with his fateful talk in Cambridge in August 1999 (see Chapters 1 and 6) and has been extraordinarily important in exploring its relation to string theory. A remarkable graduate student, Justin Khoury, was the first to join the team and played a pivotal role in developing the physical principles underlying the first colliding brane models. Nathan Seiberg collaborated on one of the early papers and has been a constant source of advice. A series of talented and creative students and postdoctoral fellows followed—Latham Boyle, Joel Erickson, Steven Gratton, Jean-Luc Lehners, Paul McFadden, Gustavo Niz, Andrew Tolley, and Daniel Wesley—all have made major contributions to unveiling new physical principles and transforming the conceptual framework into a concrete, mathematical model. Many other theorists have contributed to the cyclic model through ideas, criticisms, and insights, including Lawrence Abbott, James

Bardeeen, Thorsten Battefeld, Martin Bojowald, Robert Branden-berger, Sean Carroll, Ben Craps, Paolo Creminelli, Anne Davis, Ruth Durrer, Pyotr Horava, Gary Horowitz, Renata Kallosh, Lev Kofman, James Lidsey, Andrei Linde, Roy Maartens, V. (Slava) Mukhanov, Sub-hod Patil, Fernando Quevedo, Lisa Randall, Alexander Vilenkin, David Wands, and Matias Zaldarriaga. We also thank the United States Department of Energy and the Particle Physics and Astronomy Research Council in the United Kingdom for supporting our re-search described in this book.

Pier Gustafson, our outstanding illustrator, worked tirelessly to turn our crude ideas into illustrations that help to give the book the relaxed feel we were seeking. We were also lucky to recruit Stewart Dickson to work with us in developing the idea of a double-Klein bot-tle into a cover image that represents symbolically an endless cyclic universe. We also thank Laura van Dam and Chad Boutin for their professional editorial advice on early versions of the manuscript.

We are most grateful to Charlie Conrad, our editor at Broadway Books, and his assistant, Jenna Thompson, for their patience, advice, and support throughout the project. We appreciate the care that copyeditor Bonnie Thompson gave to the manuscript, and we thank Angela Glenn for painstakingly incorporating the edits into the final draft. Finally, we thank Katinka Matson and John Brockman for their help throughout, beginning with a sketchy concept, helping us for-mulate it into a book proposal, finding Charlie Conrad and Broadway Books to work with us, and providing many opportunities to bring the book and its ideas to the attention of the public.

Bardeeen, Thorsten Battefeld, Martin Bojowald, Robert Branden-berger, Sean Carroll, Ben Craps, Paolo Creminelli, Anne Davis, Ruth Durrer, Pyotr Horava, Gary Horowitz, Renata Kallosh, Lev Kofman, James Lidsey, Andrei Linde, Roy Maartens, V. (Slava) Mukhanov, Sub-hod Patil, Fernando Quevedo, Lisa Randall, Alexander Vilenkin, David Wands, and Matias Zaldarriaga. We also thank the United States Department of Energy and the Particle Physics and Astronomy Research Council in the United Kingdom for supporting our re-search described in this book.

Pier Gustafson, our outstanding illustrator, worked tirelessly to turn our crude ideas into illustrations that help to give the book the relaxed feel we were seeking. We were also lucky to recruit Stewart Dickson to work with us in developing the idea of a double-Klein bot-tle into a cover image that represents symbolically an endless cyclic universe. We also thank Laura van Dam and Chad Boutin for their professional editorial advice on early versions of the manuscript.

We are most grateful to Charlie Conrad, our editor at Broadway Books, and his assistant, Jenna Thompson, for their patience, advice, and support throughout the project. We appreciate the care that copyeditor Bonnie Thompson gave to the manuscript, and we thank Angela Glenn for painstakingly incorporating the edits into the final draft. Finally, we thank Katinka Matson and John Brockman for their help throughout, beginning with a sketchy concept, helping us for-mulate it into a book proposal, finding Charlie Conrad and Broadway Books to work with us, and providing many opportunities to bring the book and its ideas to the attention of the public.

> There is a theory which states that if ever anyone discovers
> exactly what the Universe is for and why it is here, it will
> instantly disappear and be replaced by something even more
> bizarre and inexplicable. There is another theory which states
> that this has already happened.
>
> —Douglas Adams, *The Restaurant at the End of the Universe*

Theories of the universe have abounded throughout human history, but the last forty years have been exceptional. A single theory, the hot big bang picture, has dominated scientific and public discourse and has even become part of the standard curriculum for school-children. Its central tenet, the idea that the universe emerged from a very hot, dense state 14 billion years ago and has been expanding and cooling ever since, has been firmly established through many independent measurements. But nearly every other feature of the theory has had to be modified. One ingredient after another—"dark matter," "inflation," "dark energy"—has been added and separately adjusted to fit the observations, and each of these adjustments has critically altered our conception of the history of the universe. Even so, the picture remains far from complete. The big bang is conjectured to be the beginning of time and space, but there is no clue as to how or why the big bang occurred. Nor is there a firm prediction about the future of the universe. Most cosmologists do not consider these flaws to be worrisome. They think that the theory will ultimately be simplified and made more complete. And perhaps they are right, Douglas Adams's joke notwithstanding.

This book concerns the emergence of a new theory of the universe, according to which our cosmic history consists of repeating cycles of evolution. Each cycle begins with a bang, but the bang is not the beginning of space or time. Rather, it is an event with a "before" and an "after" that can be described by the laws of physics. Each cycle influences the next. The events that occurred before the last bang shaped the large-scale structure of the universe observed today, and the events that are occurring today will determine the structure of the universe in the cycle to come. Perhaps space and time sprang into being many cycles ago, but it is also possible that they are literally "endless."

In this new, more integrated picture, the components that had to be added one by one to the conventional picture are either jettisoned, as in the case of inflation, or become essential, interwoven elements of the machinery that keeps the universe cycling. Most remarkably, this new theory of a cyclic universe is able to match all current astronomical observations with the same accuracy as the modified big bang picture, and it may explain some aspects of the universe that the big bang picture cannot.

To be sure, the concept of an endless universe is still in its infancy. Most cosmologists continue to accept the conventional theory, and some of them might even question the wisdom of writing a book on such a new and unproved idea. But, in our opinion, the issues that have been raised by this theory may so strongly affect the way one views the universe and humanity's place in it that they deserve to be aired even as the debate is ongoing. Furthermore, writing an account now makes it possible to offer a behind-the-scenes look at how science really works. Most books on science are written after the outcome is certain and only one theory has survived. Here there is a story within the story: a rare opportunity to capture science at an uncertain but pivotal moment, when a nascent idea has just burst on the scene to challenge an established view and the outcome is not yet known.

In a similar vein, this book provides a personal perspective: how we, the authors, were drawn to cosmology, how we were influenced by existing ideas, and how we became involved in both establishing the currently favored big bang picture and, later, creating a radical alternative.

In place of the more precise language of mathematics that physicists normally use, we have done our best to convey the concepts through stories, analogies, informal sketches, and anecdotes. We've tried to give a balanced perspective but have hidden neither our enthusiasm for the cyclic concept nor our growing concerns with the conventional picture. We regret that in trying to describe so many exciting ideas in a short book, we've been unable to give proper credit to all the deserving scientists whose work laid the foundation for the topics discussed, but hope our readers and colleagues will forgive this inevitable limitation.

The ideas, after all, are the central characters of our story. At issue are pivotal questions about the origin, evolution, and future of the universe and the forces that shaped the cosmos. And the answers, as you will discover, have extraordinary implications, not only for cosmology and physics but also for the nature of science itself and what is ultimately knowable.

endless
UNIVERSE

chapter one

2001

He was moving through a new order of creation of which few
men ever dreamed. Beyond the realms of sea and land and air and
space lay the realms of fire, which he alone had been privileged to
glimpse. It was much too much to expect that he would also
understand.

—Arthur C. Clarke, *2001: A Space Odyssey*

Two boys sit in darkened cinemas, one in London and one in Miami,
set to watch Stanley Kubrick's movie *2001: A Space Odyssey*. It is 1968, a
year of worldwide conflict and turmoil: Vietnam, the arms race, political assassinations, student protests, and rebellions. But all this is
forgotten as the film sweeps the boys along in a glorious tale of science, space, and the future.

The boy in Miami witnessed firsthand the awesome power of
technology to annihilate or inspire. Six years earlier, from his home
near Homestead Air Force Base, he watched missiles being prepared

for a strike on Cuba, knowing that his family and community would be obliterated if the looming crisis led to a nuclear exchange. Then, as the crisis subsided, he became galvanized by John F. Kennedy's promise to send a man to the Moon by the end of the decade. He emerged from these early experiences optimistic about the power of technology to improve the future and fascinated by all things scientific. He kept logbooks of every manned mission and traveled often to Cape Canaveral to observe the launches. He turned the family garage into a laboratory with large stocks of chemicals and biological specimens. And he headed to the Everglades at night, avoiding the city lights and fending off mosquitoes, to take a peek at the heavens through his telescope.

The boy in London was a refugee from South Africa, where his parents had been imprisoned for resisting the oppressive apartheid regime. But he too was optimistic, having seen the determination of people like Nelson Mandela to build a better future. Upon his parents' release, the family had left South Africa for Kenya and then Tanzania, new countries full of natural wonders—the Serengeti's wild animals and the Olduvai Gorge, home of the earliest humans. Under the hot African sun the boy had learned mathematics and science from spirited young teachers. He'd built electric motors, made explosions, and watched ant lions for hours. In 1968 his family had moved to England for the sake of the children's education, arriving in time to watch the *Apollo* moon landings on TV.

As young children, both boys had acquired their passion for science from their fathers. Each night, the father in America told stories to his little boy of Marie Curie, Louis Pasteur, and other great discoverers. The father in Africa patiently explained the Pythagorean theorem and spoke of the great achievements of ancient Greek science. Their words were like water on seeds, feeding insatiable curiosities. How does the world work? How did it start out? Where is it headed?

The boys asked the same questions that have gripped people from every society, every culture, every religion, and every continent since civilization began.

Kubrick's film speaks of a time in the foreseeable future when people will devote their skills and resources to uncovering the secrets of the universe. A space mission is dispatched to investigate a powerful signal emanating from one of Jupiter's moons. Technology, in the form of the computer HAL, threatens to end the mission, but human ingenuity and adaptability win out. A lone surviving astronaut arrives to find a giant monolith, appearing like a solid rock two thousand feet high. As he approaches, he realizes that it's actually the opening of an infinite shaft, drawing him into a transdimensional trip through hyperspace and revealing the creation and the future of the universe. Watching the film, neither boy realizes how prophetic this story might be.

A Real Space Odyssey

Fast-forward to the real 2001: rather than a lone astronaut, a worldwide community of cosmologists engaged in an intense effort to understand the beginning of the universe. The two of us, now grown, are thrilled to be among them. The boy in Miami, Paul Steinhardt, is now a professor of physics at Princeton University. The boy in London, Neil Turok, is a professor of mathematical physics at Cambridge University in England. Each of us, following his own path, has pursued his dream of becoming an explorer of the universe, albeit with paper and pencil instead of a rocketship. Three years have passed since the two of us joined forces on a risky venture to investigate a new, transdimensional view of space and time that challenges the conventional history of the universe.

Cosmologists celebrate 2001 as the year the U.S. National Aeronautics and Space Administration (NASA) launched a satellite mission from Cape Canaveral to investigate not the black monolith of Kubrick's film but a thin, dark layer of space at the outermost edge of the visible universe. The mission is called WMAP (pronounced "W-map,"), which stands for Wilkinson Microwave Anisotropy Probe. On board is a bank of highly sensitive detectors designed to gather some of the ancient light emitted from the dark layer nearly 14 billion years ago, at a time when the first atoms were just beginning to form. Every 2.2 minutes, the satellite spins once around its axis, and every hour the axis itself traces out a circle. From the combination of motions, light from a narrow ring on the sky is collected. Over the course of six months, the entire satellite keeps shifting, until the detectors have covered the entire sky. The sequence will be repeated every six months until enough light has been gathered to make a detailed portrait of the infant universe. (WMAP is a follow-up to the pioneering NASA satellite launched in 1989 called COBE, the Cosmic Background Explorer, which had made an initial low-resolution image of the early universe; in 2006, the leaders of the COBE team, John Mather at the NASA Goddard Space Flight Center and George Smoot at the University of California at Berkeley, were awarded the Nobel Prize in Physics.)

Nineteen months after the WMAP launch, in February 2003, mission head Charles Bennett and his team had collected and analyzed sufficient light to announce their initial findings at NASA's Washington headquarters, in a press conference broadcast throughout the world. One of us watched in an auditorium at Princeton University, overflowing with what seemed like everyone in town, from mailroom clerks to middle-school students, drawn by rumors of a great new discovery. The other was in a similarly packed lecture room in Cambridge, England. The sense of anticipation was tremendous, each crowd aware that its understanding of the origin

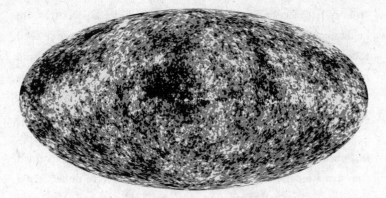

and evolution of the universe would hinge on what the WMAP team had found.

At last, Bennett and his team unveiled the image that had emerged after a yearlong exposure. Just like the fictional astronaut peering into the monolith, the WMAP satellite had gazed into the primordial layer and obtained the first clear view of the infant universe. What the greatest thinkers in history—from Plato to Newton to Einstein—could only speculate about was suddenly there for all to see, bringing humanity closer to answering the ultimate question: *Where did it all come from?*

At the end of the broadcast, world-renowned astrophysicist John Bahcall summarized the sentiments of the scientists watching: "Every astronomer will remember where he or she was when they first heard the WMAP results. For cosmology, the announcement today represents a rite of passage from speculation to precision science." Bahcall's point was that not only are the measurements marvelously accurate, but they are also in astonishing agreement with what cosmologists had been expecting.

By the time of the WMAP announcement, most scientists had come to accept a cosmological theory known as the *inflationary model of the universe*. In scientific discussions, "model" is often used to mean "theory," especially cases where the idea includes aspects that are qualitative or incomplete. The inflationary model, as the term is used today, refers to a combination of three concepts: the hot big bang model, developed in the early twentieth century; the inflation mechanism, introduced in the 1980s; and the dark energy hypothesis, added in the 1990s.

In this picture, the big bang itself is not explained. It is simply imagined that space and time emerged somehow. Next, it is assumed that just after the bang, a small region of the universe underwent a dramatic process called inflation, during which it expanded a googol (10^{100}) times or more within a billionth of a billionth of a trillionth (10^{-30}) of a second. Once this period of inflation ended, the energy causing the inflation was transformed into a dense gas of hot radiation. The gas cooled and the expansion slowed, allowing atoms and molecules to clump into galaxies and stars. This picture of an inflationary universe was originally conceived in the 1980s and is now presented in many textbooks. However, recent astronomical discoveries have led to a major amendment to the story—that 9 billion years after the big bang, a mysterious force called dark energy took over and started to accelerate the expansion again. In the standard picture, the expansion of the universe will accelerate forever, turning all of space into a vast and nearly perfect vacuum.

Both of us had been cosmologists for over two decades by the time of the WMAP announcement, and each had played a part in building the case for the leading view of the universe. In the 1980s, Paul was one of the architects of the original inflationary theory. A decade later, he and his Princeton University colleague Jeremiah Ostriker were among the first to incorporate dark energy into the big bang model. They showed that, assuming a particular mixture of matter and dark energy today, it is possible to tie together the leading ideas about the early and late history of the universe in a way consistent with all the available astronomical evidence. Neil was a leader in exploring, testing, and ruling out numerous competing notions. By showing how these alternatives failed, he helped to build the current consensus. He also predicted, on the basis of the inflationary model, a key feature of the pattern of the ancient light, which WMAP's portrait of the infant universe would later confirm.

As the two of us watched the WMAP press conference on our respective sides of the Atlantic, we were enthralled by the achievement. We both knew Bennett and most of the WMAP team personally and were overjoyed by their success. We took pride in the fact that the inflationary model, to which we had each contributed, had scored a major victory. In addition to WMAP, the model now fits an enormous range of measurements—the clustering of galaxies, the distribution of infrared radiation and X-rays, the expansion rate of the universe and its age, and the abundances of the elements—to within 10 percent or better. To have a theory that can so accurately describe events occurring billions of years ago is a stunning success; the best forecasting models cannot describe tomorrow's weather with nearly as much certainty. Fortunately, compared to the Earth's atmosphere, the conditions in the early universe are uncomplicated, and the physical laws that govern them are remarkably simple to analyze.

Yet even as we enjoyed this great step forward, we had misgivings about how the results were being portrayed as a final proof of the inflationary model. Certainly, the precise agreement between theory and observation was impressive. It was very tempting to proclaim that the big questions in cosmology were now answered. But was it really time to declare victory?

Cosmology, the study of the origin and evolution of the universe, has some unique limitations that call for a high degree of caution. Scientists cannot perform direct experiments on the universe, and they cannot travel back in time. The best they can do is gather indirect information about the history of the universe through painstaking observations of distant objects that emitted their light a long time ago and try to piece together a logical account. But the evidence is uneven, with highly detailed information about some epochs and little or no information about others. Even if one story fits all the available evidence well, there is always the possibility that another story might fit just as well, or better. Sometimes, as with Clarke and Kubrick's astronaut, a closer look will reveal that the original idea is wrong. Just as the giant monolith proved different when viewed from up close, more precise snapshots of the embryonic universe could, in the foreseeable future, lead to an entirely different explanation of the origin and evolution of the cosmos.

The chances of a dramatic shift in perspective did not seem so far-fetched. We were keenly aware that the inflationary model was not the complete and convincing picture it was sometimes portrayed to be. A number of flaws and untested predictions remained. More important, because of our own recent work, we knew of another possible explanation for the WMAP findings, one every bit as accurate as the inflationary model but based on a very different version of cosmic history.

This book will describe this more ambitious alternative, known as the *cyclic model*. According to this picture, the big bang is not the beginning of space and time but, rather, an event that is, in principle, fully describable using physical laws. Nor does the big bang happen only once. Instead, the universe undergoes cycles of evolution. In each cycle, a big bang creates hot matter and radiation, which expand and cool to form the galaxies and stars observed today. Then the expansion of the universe speeds up, causing the matter to become so spread out that space approaches a nearly perfect vacuum. Finally, after a trillion years or so, a new big bang occurs and the cycle begins anew. The events that created the large-scale structure of the universe today occurred a full cycle ago, before the last big bang.

The cyclic model accounts for the WMAP results and all other current astronomical observations with the same accuracy as the inflationary model, but the interpretation it offers is drastically different. From the cyclic view, the WMAP image is nearly as strange as Clarke's monolith. The image literally takes all of us on a transdimensional journey to view events ranging from before the big bang to the distant future.

Flaws Too Important to Ignore

In recent years many experts have downplayed the flaws of the inflationary model. Some have gone so far as to describe it as "proved," even though it has not reached anywhere near the convincing level of, say, Einstein's theory of relativity or Maxwell's theory of electricity and magnetism. While some aspects of the inflationary model have been fleshed out and thoroughly tested, only bare conceptual outlines exist for the remainder. These problems have been evident in virtually every public discussion of the inflationary model we have witnessed. Invariably, the audience, whether it consisted of trained scientists or laypeople, pointed out the same two disturbing features.

First, they say, the inflationary model appears contrived, a patchwork of disconnected ingredients that have been added one by one to fit the observations. The first ingredient is ordinary visible matter, composed of atoms and their subatomic components. A second ingredient is dark matter, an invisible entity that surrounds galaxies and accounts for most of their mass. Yet a third ingredient is dark energy, a completely different invisible substance that is spread uniformly across space and that produces an antigravity force causing the expansion of the universe to accelerate. To match the universe as it is seen today, all three components must exist in a particular precise combination. In addition, to explain the past evolution of the universe, the inflationary model requires still another ingredient, known as inflationary energy. This ingredient is needed to drive the explosive expansion of the universe for a few instants after the big bang, and then it must decay away in order not to interfere with the subsequent evolution of the universe.

As the inflationary model has matured and more components have had to be added, no one has been able to explain how or if the components might be related. Each new ingredient has required its own awkward adjustment to fit with the rest of the model. Scientists are trained to be suspicious of such procedures. They seek simple, all-encompassing explanations for the fundamental mechanisms underlying nature, like Newton's three laws of motion, which summarize everything known about motion and forces before special relativity; or the four-letter language in which the genetic code of life is written; or Schrödinger's equation in quantum theory, which describes with one stroke the structure of every type of atom and molecule. None of these great advances was made by adding awkwardly tweaked components to a preexisting idea. On the contrary, each breakthrough was inspired by dissatisfaction with such an approach. In the same way, every new ingredient and every special adjustment in the inflationary model may be a further indication that it is not the final answer.

The best hope for finding a more compelling explanation lies in the realm of fundamental physics: the attempt to describe all the forces and particles of nature in a single, unified theory. Dramatic progress toward such a unified theory has been made over the past two decades. But so far there is no clear indication how the complex mix of dark matter, dark energy, and inflationary energy required by the inflationary model might emerge from such a unified theory.

The contrived nature of the inflationary model and its failure, so far, to connect with fundamental physics in a simple way provide powerful reasons for seriously considering alternatives. The second common objection and, for many, the most disturbing feature of the inflationary model by far is the idea that time has a "beginning." How did the universe start, if there was nothing before it? The notion sounds contradictory, and maybe even nonsensical.

Philosophers have grappled with the issue for thousands of years without making much progress. Some people appeal to religious faith for arguments supporting the creation of something from nothing. Even here, however, a careful reading of the original texts reveals a certain ambiguity, as will be discussed in chapter 8. For example, in the Judeo-Christian tradition, the opening lines in the Book of Genesis do not make clear if they are describing the creation of the Earth and Sun or of the entire universe. In other places, the word used for "creation" refers to molding the world from *preexisting* formless material. Many leading biblical scholars interpret the scriptures as suggesting the possibility of previous acts of creation.

The known laws of physics are not of much help, either, since they describe how things evolve in time and not how time can begin. A number of physicists have suggested imaginative modifications of the known laws to prescribe how space and time "emerged" from a primordial state or from nothing at all. However, even if one accepts these proposals for how space and time sprang into being, none of them provides a convincing explanation for why the universe started out with a high concentration of the kind of energy required for the inflationary picture.

If the inflationary model is flawed and scientists cannot understand how inflation began, what can be done? The questions of whether the universe had a beginning or not and how it will evolve in the future are too important to be ignored. Their resolution will not only settle questions about the history of the universe but will also illuminate the fundamental laws of physics. The conditions of high temperature and high density in the early universe lay bare the elemental forces that created the galaxies and stars and that continue to govern the universe today. Given the ramifications, there is simply no choice but to press on, either addressing the flaws of the inflationary model or finding a better idea.

Dreams of a Better Theory

History should encourage one to think boldly. Several times in the last century, cosmologists converged on what they believed to be the true model of the universe, only to discover that, with new observations and advances in theoretical physics, they had to abandon it in favor of a radically new idea.

Before Einstein, many astronomers had concluded that only a single, isolated galaxy of stars, the Milky Way, existed, surrounded by an infinite expanse of empty space. When Einstein developed his own model of the cosmos based on his new theory of gravity, known as general relativity, he assumed a different picture in which the matter in the universe is spread uniformly throughout space and the universe has existed in its present state for all eternity, neither expanding nor contracting. Over the next decade, the astronomer Edwin Hubble showed that the astronomers' picture was wrong. In 1923, Hubble showed that stars are not spread out uniformly, but are instead clumped into galaxies that are spread throughout the universe far beyond the boundaries of the Milky Way. In 1929, Hubble further proved that, contrary to Einstein's concept, the state of the universe is changing because the galaxies are all moving apart from one another and space is expanding. In the 1950s, Fred Hoyle, Hermann Bondi, and Thomas Gold, working together at the University of Cambridge, revived the idea of an unchanging universe, convincing many astronomers that matter is created at just the right rate to balance the expansion of the universe and keep it in a steady state for all time. But the 1963 discovery of the cosmic background radiation by Bell Laboratory astronomers Arno Penzias and Robert Wilson shattered the steady-state model. The radiation, which comes from the early universe, is direct evidence that there was once a period when the temperature and density were much greater than today.

The discovery of the cosmic background radiation immediately convinced almost all cosmologists of the hot big bang model, first proposed by Russian mathematical physicist Alexander Friedmann and Belgian cosmologist Georges Lemaître in the 1920s, based on Einstein's theory of gravity, and developed by George Gamow and collaborators at George Washington University in the 1940s. Their model was simple compared to today's inflationary model: it assumed a universe containing only ordinary matter composed of atoms and their components. By the 1980s, cosmologists had been forced to introduce dark matter and inflation to explain the observed motions within galaxies and galaxy clusters and the formation of large-scale structure in the universe. Then, in the early 1990s, just as most cosmologists were becoming convinced that the inflationary big bang model was the answer, they were shocked to discover the existence of dark energy and cosmic acceleration in today's universe.

Although this history suggests that radical changes in cosmology are possible, whenever either of us imagined constructing an alternative model that would address the flaws in the inflationary picture, the prospect seemed daunting. Whatever the new concept might be, it could not be a simple variant of the old ones, since those had already been considered, tested, and eliminated. An alternative model would require completely novel features. But the features had to be well justified, scientifically sensible, and mathematically consistent. A new model that tries to embrace the entire history of the universe would probably not emerge fully developed. More likely, it would first appear as a tenuous framework mixing a few solid ideas with a lot of bold speculations. Building firm foundations would probably require several years. And at any time a fatal flaw might appear—due to either a mathematical surprise or a new observation—and topple the whole idea.

These thoughts played in the backs of our minds for years. Neil was often outspoken about them; Paul tended to keep them to himself. For over twenty years, we had worked on different problems with many different collaborators, but not with each other. Yet, based on casual remarks exchanged during informal encounters over the years, both of us sensed a common willingness to consider radical alternatives, if the opportunity ever arose.

On August 19, 1999, the two of us sat on opposite sides of a lecture room at the Isaac Newton Institute for Mathematical Sciences in Cambridge, listening to a lecture by Burt Ovrut from the University of Pennsylvania. Suddenly, we both were struck by the same insight.

Ovrut was describing string theory, currently the most promising attempt toward a unified theory of the fundamental forces that govern all physical processes. More specifically, he was reviewing how a particular form of string theory naturally leads to a new theory of elementary particles, according to an idea suggested by Edward Witten, one of the leading pioneers of string theory, and Petr Hořava, at the Institute for Advanced Study in Princeton. In this version of string theory, the ordinary three-dimensional "world"—everything in the universe that can be touched, felt, or seen—is separated from another inaccessible three-dimensional "world" by a tiny gap along a fourth dimension that cannot be experienced. Atoms and light can move in the height, length, and depth of this world but are forbidden by the laws of string theory from moving into the extra dimension. The other world has its own kinds of matter and light, which cannot travel through the extra dimension either. Thus the two worlds are totally disconnected from each other except for one factor: they interact through the force of gravity. Ovrut, working with the assumption that the worlds lie at a fixed, minuscule distance from one another along the extra dimension, described how this bold new idea explained many of the detailed features of elementary particles.

After the lecture, both of us converged on Ovrut from different directions. No one can recall who blurted it out first, but one of us asked, "Can't these worlds move along the extra dimension? And, if so, is it possible that the big bang is nothing more than a collision between these two worlds?" It became immediately clear that we had been struck with the same vision. Ovrut's compelling conceptual picture suggested that the big bang might not be the "beginning" of the universe after all, but instead a physically explicable event with a "before" and an "after." Furthermore, if there was no inflation to spread apart and dilute the matter and structures produced at the collision and one could observe them today, there might even be direct observational evidence of events that had occurred before the big bang.

This was the first of many "Aha!" moments we were to experience over the next few years, on an intellectual roller-coaster ride full of thrills, dips, and more thrills. From rudimentary beginnings, a new vision of the history of the universe gradually emerged that eventually became the cyclic model.

At the conceptual level, the cyclic model differs from the consensus picture in three key respects. First, the big bang is transformed from a singular beginning of space and time into a collision between worlds, just as we had proposed after Ovrut's talk. Second, the big bang is not a one-time occurrence. The worlds are drawn together and collide at regular intervals of about a trillion years. Each bang creates new hot matter and radiation and initiates a new period of cosmic expansion, leading to the formation of new galaxies, stars, planets, and life. Finally, what happens before the big bang cannot be ignored. The spatial arrangement of galaxies and the patterns observed in the cosmic background radiation today are set in place by events that took place a cycle ago. Likewise, events taking place today are setting the scene for the pattern of galaxies and radiation in the cycle to come.

Despite the profound differences, the cyclic model reproduces every success of the inflationary model to date. The cyclic model has now developed enough to enter the scientific fray and challenge the inflationary model. An exciting debate has begun.

A Cosmic Competition

This book presents the two contrasting views of the cosmos. The story begins with the period of cosmic history on which the two models agree, between the one-second mark after the last big bang and the present time. A wide range of astronomical observations, especially those made during the last decade, leave little room for doubt about what happened during this epoch. But when one asks what happened before the one-second mark and what will occur in the future, the cyclic model and the inflationary model provide dramatically different answers.

A critical aspect in judging the two models is how they relate to the fundamental laws of physics. The inflationary model arose from the view that matter is composed of indivisible particles, interacting through forces. The cyclic model was motivated by the revolutionary ideas of string theory, the new approach to fundamental physics that has grabbed the attention of leading theorists in recent decades. The basic tenets of string theory—that matter is composed of vibrating stringlike objects and that space contains extra hidden dimensions— imply a new geometrical view of the universe. That vision led directly to the formulation of the cyclic picture.

It is too early to say how the debate will be resolved. The ultimate arbiter will be nature, as it is for all scientific debates. Scientists are already planning experiments that could, over the next decade or so, determine which, if either, cosmological model is correct.

As the scientific community is beginning to accumulate the observations that could settle the issue of where the universe came from and where it is going, there is still time for everyone to enjoy speculating about how the answer will affect many other profound cosmological questions: Are the laws of physics the same everywhere? Do space and time last forever? Is there only one universe? What will ultimately happen to the matter and light observed today?

This tale captures science at its most exciting moment, when the outcome of a remarkably important debate is still in doubt. The book has been written with the fervent hope that the reader will be swept up by the issues, amazed by the technology that enables scientists to investigate them, and thankful for the good fortune of being alive at a time when such grand ideas are being explored.

chapter two

Act Two

You can't make a man unsee what he has seen.

—Bertolt Brecht, *The Life of Galileo*

The history of the universe can be compared to a play in which the actors—matter and radiation, stars and galaxies—dance across the cosmic stage according to a script set by the laws of physics. The challenge for the cosmologist is to figure out the story line after arriving at the show 14 billion years too late, long past the crucial opening scenes.

Observations of nearby galaxies and stars provide an accurate picture of the present scene. By gathering light emitted long ago from more distant objects and applying the physical principles learned and tested on Earth, astronomers have been able to reconstruct more and more of what happened in the past. The epoch we call Act Two, which began just one second after the big bang and continues to the present day, is the period of cosmic history that is best understood. During this nearly 14-billion-year span, the universe expanded more than a billion times in size, and the hot primordial gas that filled the infant universe cooled to less than a billionth of its initial temperature, condensing into structures of increasing mass and complexity: the first atomic nuclei, the first atoms and molecules, and, ultimately, the first planets, stars, galaxies, galaxy clusters, and superclusters.

One might imagine that, armed with this detailed knowledge of Act Two, scientists could straightforwardly determine what happened before or what will happen next. The big surprise is that this is not the case. The inflationary and cyclic models *both* incorporate Act Two, but they sandwich it between completely different first and third acts. In the case of the cyclic model, Act Three is not even the final act: the plot eventually leads to a new Act One, in which the story begins all over again. To appreciate how two radically different views of the history of the universe can both be consistent with the plethora of observations available today, one must first understand what is and is not known about Act Two.

The Cosmic Sphere

We live on a minor planet orbiting an ordinary star, the Sun, one of a hundred billion stars making up the Milky Way galaxy, which is visible on a clear night as a smear of light stretching across the sky. Beyond the Milky Way, powerful telescopes can see a hundred billion other galaxies, spread through space in a complex, hierarchical pattern of galaxy clusters and superclusters.

The key to discovering the structure of the universe, the way it evolved, and where it is headed is to gather light from distant sources. Light travels through empty space at a finite speed, 186,000 miles per second, so the light collected today from a distant source must have been emitted long ago. This is why astronomical images show objects as they once were, rather than as they are now. Based on this simple notion, powerful telescopes can be viewed as time machines to study the evolution of the universe from very early times up to today.

To emphasize the point, astronomers label distances according to the amount of time required for light to travel from the source to the Earth. For example, the average Earth-Moon distance is called 1.282 "light-seconds" because light takes 1.282 seconds, on average, to travel from the Moon to the Earth. By the same reasoning, Pluto is five and a half light-hours away. The nearest star, Proxima Centauri, is 4.2 light-years away.

The Milky Way's cosmic neighbor, the Andromeda Galaxy, lies 2.9 million light-years from the Earth; the light received from its stars today was emitted before the earliest humans roamed the Earth. At a distance of 52 million light-years lies the giant Virgo cluster, consisting of well over a thousand galaxies. The most distant cluster of galaxies ever observed is 10 billion light-years away. And the farthest

region ever observed is 13.7 billion light-years away, the layer of space portrayed by the WMAP satellite. The radiation collected to form the WMAP image, known as cosmic background radiation, was emitted from this layer only 380,000 years after the big bang. Prior to that moment, the universe was too hot for atoms to exist. Instead, atoms were broken apart into a gas of charged atomic nuclei and electrons called *plasma*. The plasma scattered light so strongly that the universe was completely opaque. After 380,000 years, the plasma cooled off enough for the nuclei and electrons to combine into a transparent gas of neutral atoms. From that moment on, light traveled freely through the universe.

One can imagine using all the astronomical observations that have ever been made to construct a scale model of the optically visible universe in the form of a filled "cosmic sphere" with the Earth at its center. The cosmic sphere represents the universe as seen from the Earth. A similar sphere could be drawn around a planet in any other galaxy. Although the fine details would be different, the average appearance of the sphere would be the same.

If each galaxy in today's universe occupied a region the size of a grain of sand, the billions of galaxies within reach of telescopes would fill a sphere a few meters across. Galaxies near the Earth are seen more or less as they are today. Distant galaxies appear as they were billions of years ago, because the light received from them had to travel for billions of years to reach us. The outermost galaxies are pictured as they were when they were forming their first stars. Some contained bright quasars, believed to be giant black holes devouring clouds of hot gas swirling about them. Beyond all the visible galaxies lies a dark shell of space, which appears to be totally devoid of stars or galaxies. In actuality, the dark shell is no different than any nearer region of space. Galaxies and stars formed there at about the same time as the galaxies and stars nearby. The dark shell only seems to be

vacant because it is so far away that it appears as it was before any galaxies formed. Finally, the outer skin of the cosmic sphere displays the WMAP image, whose light has taken almost 13.7 billion years to reach us. This light was emitted as the atomic nuclei and electrons in the hot plasma combined to form the first atoms, long before the formation of galaxies.

This concept of building a scale model of the universe is not too far from reality. With the help of powerful new technologies, astronomers have been gathering information needed to complete the cosmic sphere. One of the biggest challenges is measuring the distances to remote galaxies. The standard approach entails comparing the brightness and colors of distant astronomical sources with those of similar objects seen nearby whose distances can be measured directly. The techniques are painstaking and time-consuming because the distant galaxies are extremely dim. A key technological breakthrough has been the invention of charge-coupled devices (CCDs), which are sensitive light detectors similar to those in digital cameras. Using a specially designed, ultralarge-format CCD camera designed and constructed by James Gunn at Princeton University and mounted on a 2.5-meter telescope at Apache Point, New Mexico, a team of astronomers from fourteen institutions have been engaged since 1998 in the Sloan Digital Sky Survey. This project has measured the distances to over 2 million galaxies in slices of space spread over a quarter of the sky in the Northern Hemisphere and extending outward a billion light-years, about a tenth of the way to the edge of the cosmic sphere. A similar survey, called 2dF (for "two-degree field"), conducted at the Anglo-Australian Observatory in central New South Wales, Australia, has mapped a large patch of sky in the Southern Hemisphere.

To see farther, cosmologists have been using the Hubble Space Telescope, the first space-based optical observatory, in orbit four hundred miles above the Earth. The Hubble telescope was not designed for this purpose. With its ability to focus on tiny patches of sky and study the fine details of individual objects, the telescope was considered to be incompatible with the kinds of broad surveys needed for cosmology. However, thanks to the foresight of Robert Williams, the director of the Space Telescope Science Institute in Baltimore, Maryland, the Hubble Space Telescope has been able to make a vital contribution.

Astronomers flood the Space Telescope Science Institute each year with proposals for using the Hubble telescope to study specific objects. One of the perks of being the institute's director is having personal control of a certain amount of telescope time. In 1995, after gathering advice from many leading astronomers, Williams made a selection that might at first sound bizarre. He chose to use a full ten days of telescope time, spanning 150 Earth orbits of the space telescope, to stare at an area that appeared to be totally blank, devoid of stars or galaxies. Williams figured that the region was ideal for obtaining an unobstructed view of what, if anything, lies beyond.

The chosen patch of the sky was a speck, roughly the size of Franklin D. Roosevelt's eye on a dime if you hold the coin out at arm's length. The patch appears blank to the human eye and to ground-based telescopes because it contains no stars or galaxies bright enough to be seen. But by adding up the light collected over ten days, the Hubble Space Telescope was able to gradually build a spectacular picture. The result, known as the Hubble Deep Field image, shows thousands of distant galaxies with a myriad of shapes and sizes, pro-

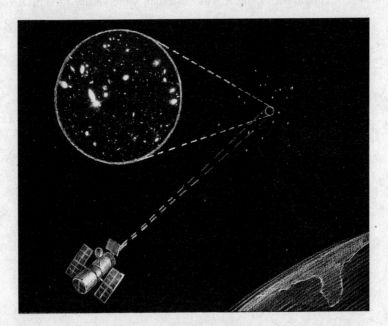

viding a direct view of the adolescent universe when the first galaxies were still forming, billions of years before they assumed the shapes observed today. The Hubble Deep Field image provides a good impression of what the cosmic sphere looks like all the way out to the most distant visible galaxies.

As we have already noted, the information needed to paint the outer skin of the cosmic sphere is being provided by the WMAP satellite. As of 2007, the satellite is orbiting the Sun a million miles from the Earth and continuing to refine its image of the sphere's outer layer. As first pointed out over thirty years ago by cosmologists P. J. E. (James) Peebles and Jer Yu at Princeton University and Rashid Sunyaev and Yakov Zel'dovich at the Moscow Institute of Applied Mathematics, the very slight temperature variations in the cosmic background radiation across the sky contain a vast amount of information about the early universe. They provide a direct picture of how the density varied from place to place. The primordial density variations were essential to the later formation of galaxies and other large-scale structures in the universe. Where the matter was slightly denser than average, gravity caused it to clump further. Eventually, it collapsed inward, forming a galaxy or a cluster of galaxies. Hence the features seen in the WMAP image are the ancient progenitors of galaxies, and of the stars and planets within them. Each of us owes our existence to early density variations like those seen in the WMAP picture, which led to the formation of the Milky Way.

The first experiment to detect variations in the temperature of the radiation across the sky was the Cosmic Background Explorer (COBE) satellite in 1992, but its resolution was too poor to depict the fine features. More than a decade later, in 2003, the WMAP satellite succeeded in producing the first all-sky, high-resolution picture. The image is packed with detailed information, which has been used to flesh out the story of Act Two. In the future, still more refined measurements of the cosmic background radiation may help to determine what happened in Act One and what is likely to occur in Act Three.

The boundary of the cosmic sphere is the layer from which the cosmic background radiation was emitted, 380,000 years after the bang. This layer is the outer limit of what can be seen using light and other forms of electromagnetic radiation, because the hot plasma that filled the universe at earlier times obscures the view of the universe at greater distances.

To learn about the universe at yet earlier times, cosmologists use indirect methods. One of the most powerful involves studying the abundances of the different light chemical elements—hydrogen, deuterium, helium, and lithium—in the universe today. The nucleus of a hydrogen atom consists of a single proton, but other atomic nuclei are made of various combinations of protons and neutrons. One second after the big bang, the universe contained free protons and neutrons, but the temperature was too high to allow heavier atomic nuclei to exist. They only formed in the ensuing few minutes, as the universe expanded and cooled, allowing protons and neutrons to stick together for the first time. The resulting abundances of deuterium, helium, and lithium depend very sensitively on the rate of expansion and the density of protons and neutrons during those early moments. By actually measuring the abundances of different atomic nuclei in clouds of primordial gas, astronomers provide cosmologists with the information needed to infer the conditions in the early universe. The measurements can be compared with computer simulations of the cosmic expansion and the nuclear processes occurring in the first few minutes after the bang. The comparison shows a striking agreement between theory and observation if the evolution of the universe is described from the first second onward by Einstein's theory of general relativity, and if the nuclear processes obey the same laws as those obtained from studies in nuclear reactors and particle accelerators today. The successful match is what gives cosmologists confidence that they can trace the detailed history of the universe as far back as the first second, which is why we set the one-second mark as the beginning of Act Two.

Ultimate Plastic™

The cosmic sphere is a very useful way of compiling astronomical observations of the region of the universe around us, but it has one major limitation: it ignores the fact that the universe is expanding. For example, the WMAP image painted on the outer layer appears to span an enormous area, a spherical surface 13.7 billion light-years in radius. However, at the time the radiation was emitted from the WMAP layer, the universe was a thousand-fold smaller than it is today. Thus the WMAP layer was just 13.7 *million* light-years in radius at that time, only a quarter of the distance between the Milky Way and the nearest galaxy cluster today.

The expansion controls the temperature, density, and composition of the universe, and the formation of astronomical structures. The realization that space can expand or contract traces back to Einstein's theory of gravity, known as the general theory of relativity. The central tenet of Einstein's theory is that space and time form an elastic substance called space-time that can stretch, contract, warp, or wiggle. The gravitational force is due to the warping effect that a massive object has on the space around it, analogous to the depression created when a bowling ball is placed on a soft, springy mattress. When other objects travel through warped space, their paths are distorted, analogous to the way a marble rolling on the mattress will swerve when it encounters the depression created by the bowling ball. As John Wheeler of Princeton University, a leading exponent of general relativity, likes to put it, "Mass tells space-time how to curve; and space-time tells mass how to move."

In fact, according to Einstein's best-known equation, $E = mc^2$, mass is just one form of energy. Wheeler's dictum remains true if the word "mass" is replaced everywhere with "energy," which can take any form whatsoever. (E stands for energy, m for mass, and c for the speed of light.)

Now imagine that all of space is filled with a nearly uniform distribution of matter, radiation, or other types of energy. Then, according to Einstein's theory, the entire universe can expand or contract, just like the overall stretching or shrinking of an elastic sheet. Ironically, Einstein himself resisted the idea of an expanding or contracting universe, even though it is a natural consequence of his own theory. For philosophical reasons, he strongly preferred a model in which the universe is static and eternal.

So committed was Einstein to his vision of a static universe that he introduced an extra form of energy specifically designed to oppose the gravitational attraction of matter. The new form of energy, which came to be known as the *cosmological constant,* has a repulsive gravitational effect, causing space to expand rather than contract. In his first model of the universe, Einstein finely adjusted the repulsive gravity of the cosmological constant to counter precisely the attractive gravity of matter. By setting the opposing influences in perfect balance, Einstein was able to construct a static model of the universe. However, this situation is contrived and unstable: unless the balance between forces is perfect, the universe either collapses or blows up.

The empirical proof that the universe is not static came ten years after Einstein's proposal. Edwin Hubble, for whom the Space Telescope is named, observed the motions of distant galaxies and found convincing evidence that they are spreading apart and that the universe is expanding. To appreciate Hubble's evidence for cosmic expansion, consider first an imaginary pocket toy, similar to the types that appeared in advertisements in old comic books. The ad might read: "*Ultimate Plastic*™: Imagine holding a universe in the palm of your hand!" The toy comes as a cube the size of a sugar lump, with handles at each corner for you and some friends to stretch apart. Pull on the handles and the cube grows in size. Keep pulling and you can make the cube as big as a room. Inside the plastic, the makers have sprinkled tiny models of galaxies that you can see spreading apart from one another as the cube expands. The model galaxies are made of a hard material so that they do not expand when the cube is stretched—only the space between them does. This was Hubble's mental image of a chunk of the expanding universe.

None of the galaxies is special: if you could shrink yourself and perch on any one of them, you would see all the other galaxies moving away from you. Furthermore, each time the sides of the cube double in length, all distances between galaxies double. Suppose the doubling takes place in 1 second. Then a galaxy that starts out 2 meters away from you winds up 4 meters away. So, it has receded at an average speed of 2 meters per second. A galaxy that is initially 3 meters away will be 6 meters away after 1 second, so it has receded faster than the first at an average speed of 3 meters per second. A galaxy that is initially 5 meters away has receded with an average speed of 5 meters per second. In other words, the farther away a galaxy is, the faster it appears to recede. Hubble found just this pattern for real galaxies; the observation that galaxies are receding from the Earth at a speed proportional to their distance is known as Hubble's law.

Not only was Hubble's conclusion important, but the methods he used are still applied today, with various improvements, to mea-

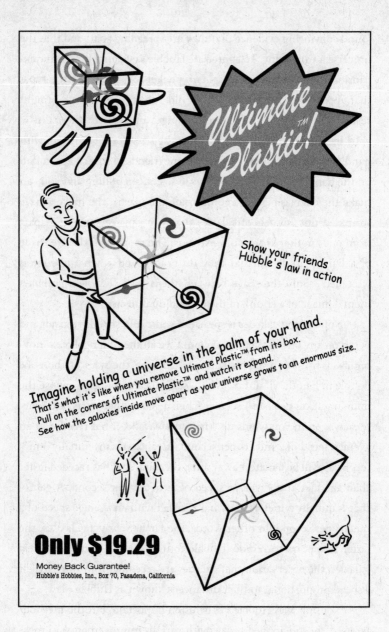

sure cosmic distances. To determine the distance to galaxies, Hubble relied on the fact that nearly all galaxies contain pulsating stars known as Cepheids, whose brightness varies with time according to a regular, repeating rhythm. In 1912, Henrietta Leavitt, working at the Harvard Observatory, observed many Cepheids in the Milky Way and showed that those pulsating with the same rhythm emit the same amount of light. A decade later, using the 100-inch Hooker telescope at the Mount Wilson Observatory, the most powerful telescope on Earth at the time, Hubble managed to observe Cepheids in other galaxies. By comparing their apparent brightness to that of Cepheids in the Milky Way pulsating at the same rate, Hubble could estimate how far away their host galaxies must be.

In order to determine how fast each galaxy is moving toward or away from us, Hubble also measured the colors emitted by the Cepheids. Each star emits particular colors, depending on the composition of its atmosphere. For example, all stars contain hot hydrogen gas, which gives off light with a specific pattern of colors. If the star is stationary with respect to the Earth, the colors observed are the same as those of hot hydrogen gas measured in the laboratory. If the star is moving away from the Earth, though, each successive crest in the light wave has to travel a greater distance to reach the Earth, and, hence, the apparent wavelength is longer than it would be if the star were at rest. A longer wavelength of light corresponds to a redder hue. Therefore, the greater the speed at which the star recedes from the Earth, the greater is the red shift. If the star is instead moving toward the Earth, the pattern of colors is shifted toward bluer hues. Astronomers refer to the color changes as red shifts and blue shifts.

If the universe were static, as Einstein had first expected, Hubble might have found the motion of galaxies to be random, just like the molecules of air in motion all around us. Instead, he found something remarkable: except for the nearest galaxies, *all the other galaxies are moving away from us.* When he plotted the speed versus the distance for each galaxy in his study, the graph showed a straight line: the speed at which a galaxy recedes is directly proportional to its distance from the Earth, just like the model galaxies in Ultimate Plastic™. Here was a strong indication that space is uniformly stretching, just like the plastic in the imaginary toy. The slope of the straight line in his plot is called the Hubble parameter. It is a measure of the rate of expansion of the universe.

Since Hubble, observers have been using light from various kinds of astronomical sources to judge the range and recessional speed of galaxies and to confirm the validity of Hubble's law at greater distances. Each type of source emanates from somewhere within a galaxy and has some distinctive feature that makes it possible to determine how much light it emits. Then its distance can be judged by comparing its apparent brightness with the apparent brightness of nearby sources of the same type, whose distance can be measured directly. An example is the Type IA supernova, a white dwarf star that accretes matter from a second star in orbit around it until it reaches a critical mass and explodes. Type IA supernovae have the advantage that they are far brighter than Cepheids and so can be used to judge the distance of extremely remote galaxies.

In our toy, the plastic expands but the model galaxies do not. Similarly, the expansion of the universe stretches the distances between well-separated galaxies but not the galaxies themselves, nor their components, including stars, planets, and people. The latter are held together by forces strong enough to resist the expansion of the universe. For example, solid objects, such as the human body, are held together by chemical bonds that are far stronger than the cosmological stretching force.

Hubble's law—that the recession speed grows in direct proportion to distance—is now established for galaxies within a few billion light-years of the Earth. Beyond that, naive notions of speed and distance need to be revised, because the recession speed as estimated from Hubble's law rises toward the speed of light. In this situation, the laws of special relativity need to be carefully applied. Nevertheless, the notion of space as Ultimate Plastic™ continues to provide the right insight about the expansion of the universe, and its effect on light and other forms of radiation, no matter how distant the source is from us. Each time the universe expands by a factor of two, the wavelength of light traveling through space also doubles. For example, the universe has stretched by a factor of a thousand since light was emitted from the WMAP layer. Today, the cosmic background radiation has a wavelength of a few millimeters and a temperature of 2.7 degrees above absolute zero (i.e., minus 270 degrees Celsius). But at the time it was emitted, the wavelength of the radiation was a thousand times shorter—a few microns, the wavelength of infrared light.

The energy and temperature of radiation grows in inverse proportion to its wavelength. Hence, cosmologists infer that the temperature of the universe was nearly three thousand degrees, or half the surface temperature of the Sun, at the time when the cosmic background radiation was emitted. This is hot enough to boil neutral atoms into charged nuclei and electrons. As the universe is traced even further back in time, the wavelength of the radiation gets smaller and the temperature grows ever higher. From what is known about how matter and energy change their character as the temperature rises, cosmologists can pinpoint many of the key transformative moments in the history of the universe.

Friedmann and the Expansion of the Universe

The discovery that the expansion (or contraction) of space is a natural outcome of general relativity theory was made by the Russian mathematical physicist Alexander Friedmann, working in St. Petersburg in 1922. Assuming a nearly uniform distribution of matter and no cosmological constant, Friedmann showed that space could not stand still. It would have to expand or contract everywhere in the same way, just like the Ultimate Plastic™ toy. Friedmann showed that in this case, Einstein's famously complicated equations of general relativity can be reduced to a simple formula, known as the *Friedmann equation*, linking the Hubble parameter to the curvature of space and the energy density of the universe, that is, the energy per cubic centimeter of space.

The curvature of space measures the deviation from the normal rules of geometry everyone learns in high school, rules dating back to Euclid. According to Euclidean geometry, two parallel lines maintain the same separation all along their paths. Equivalently, one can imagine sending off two parallel laser beams in any direction in three-dimensional space, and the two beams will never cross or diverge. Cosmologists use the term *flat* to describe space with this property. But Einstein's theory of gravity does not require that space be flat. Space can have a positive curvature, so that it bends back on itself like the surface of a sphere. In a positively curved space, two initially parallel light beams approach and intersect, just as the lines of longitude on the surface of a globe meet at the north and south poles. Cosmologists often refer to this case as a *closed* universe. Conversely, if the curvature is negative, space opens outward and two parallel light beams diverge. This case is often called an *open* universe. The curvature is an important factor in determining the long-term future of the universe. For example, if the universe contains only matter and radiation, an open or flat universe expands forever but a closed universe expands for only a finite time before contracting into a big crunch.

Because of the importance of the curvature, cosmologists tried for decades to determine whether the universe is flat, open, or closed. Finally, the WMAP measurements settled the issue, showing conclusively that the curvature is negligible and space is very nearly flat on large scales. But nothing in Einstein's general theory of relativity or Friedmann's analysis explains why there should be no curvature. The search for a natural explanation of this fundamental feature of the universe is one of the prime motivations for the inflationary and cyclic pictures, as will be explained later in the book.

In a spatially flat universe, Friedmann's relation becomes very simple: the expansion rate of the universe is proportional to the square root of the energy density. The expansion of the universe in its turn determines how the temperature and energy density change. Appropriating Wheeler's dictum, one might say, "The energy density determines how fast space expands, and the expansion rate determines how fast the energy density falls."

All forms of energy must be counted, but each form of energy responds to the expansion differently. For matter, the energy density decreases because the volume grows and the matter spreads out. For radiation, there is an additional effect: not only is the density diluted but the expansion of the universe stretches the wavelength of each individual light wave, thereby depleting its energy. Thus the energy density of radiation falls even faster than the energy density of matter. For dark energy, the energy density remains nearly constant as the universe expands. Because of these differences, the relative proportions of radiation, matter, and dark energy change over time. At early times, radiation dominates the energy density of the universe.

Later on, the density of radiation falls below that of matter. Finally, in the late universe, the dark energy dominates the density of the universe as both matter and radiation are diluted away. This explains why Act Two is divided into three scenes: the radiation-, matter-, and dark energy–dominated epochs.

The transition from radiation to matter dominance about 75,000 years after the big bang was a critically important event in cosmic history. Structure forms in the universe from slight variations in density, such as those observed in the WMAP image, when matter clumps gravitationally around the regions with higher than average density. As long as radiation dominated the universe, though, the expansion rate was too rapid for structure to form. Only when matter came to dominate did the expansion slow enough to allow regions of higher density to draw together to form galaxies, within which stars and planets were born. The formation of the solar system and the Earth, therefore, depended crucially on this transition from radiation to matter domination.

The formation of the Earth also depended critically on the sequence of events that occurred during the radiation-dominated period that preceded matter dominance. Just a few seconds after the big bang, the temperature of the universe was a billion degrees, a hundred times hotter than the center of the Sun. This is the period discussed earlier when protons and neutrons first fused into helium and other nuclei, those fusion reactions marking the beginning of Act Two. Later on, the primordial helium underwent further fusion reactions in stars that produced carbon, oxygen, and yet heavier elements. Most of the heavy elements that make up the Earth were formed in exploding stars called supernovae, and then scattered through space. Without the primordial helium emerging from the early universe, however, there would not have been enough heavier elements to form the iron cores of terrestrial planets, like the Earth, or the molecules on which life depends.

Cosmologists would like to extrapolate further back in time to discover how the matter and radiation themselves originated, but this is problematic. Extrapolated backward just one second before the start of Act Two, the Friedmann relation implies that the scale of the universe shrinks to zero and the energy density and temperature grow to be infinite. This is the moment known as the big bang, also referred to as the *initial singularity*. Mathematicians use the term "singularity" to indicate that equations are failing. The big bang is referred to as the initial singularity because Einstein's equations of general relativity break down when the temperature and energy density become infinite, as Einstein himself recognized, and their description of the expansion of the universe ceases to be valid.

When equations develop singularities, physicists normally interpret this to mean that the equations are being extended into a regime where they can no longer be trusted, and that the laws of physics they were using must be replaced by improved ones capable of making sense of the situation. For example, the flow of air around the wing of an airplane can usually be accurately modeled by treating air as a smooth fluid. But if the plane exceeds the speed of sound, a shock wave forms and the fluid equations predict that the flow develops singularities, with the density of the air becoming infinite at some locations. In this situation, the equations can no longer be used. The remedy is an improved theory of shocks in which the air is described more precisely as a collection of molecules bouncing off one another. String and M theory, described later, are attempts to improve Einstein's theory of gravity that represent space-time in terms of new, more fundamental microscopic entities. Most cosmologists hope that such an improved description of gravity will ultimately tackle the problem of the cosmic singularity, and some positive indications of this will be explained in subsequent chapters.

Today, many cosmologists interpret the singularity in Friedmann's equation as signifying the beginning of space and time, and most textbooks present this notion as certain. However, this is an assumption, not a proven fact. The cyclic model of the universe challenges this point of view, suggesting that the big bang was not the beginning of time but rather a violent transition between two stages of cosmic evolution, with a "before" and an "after." In fact, according to the cyclic model, the big bang in our past was caused by a strange substance that is now starting to take over the universe and that will eventually lead to the next big bang in our future, and the one after that.

The strange substance is one of two shady characters that astronomers had to introduce in order to make sense of Act Two. The original hot big bang model, developed in the first half of the twentieth century, included only ordinary atomic matter and radiation. As it turned out, two additional components were initially missed by astronomers but were lurking in the background all along, exerting a profound, controlling influence on everything we now see, on everything that has transpired in Act Two, and on everything that will happen in Act Three.

Two Shady Characters

The two new characters neither emit nor absorb light, so they are hard to find. Their composition is unknown. They have never been produced in a laboratory. They are not composed of protons, neutrons, or electrons, or any of the many more exotic elementary par ticles that have so far been produced in the high-energy collisions that take place in particle accelerators. Where they came from or what they will do in the future is a mystery. But what is certain is that, together, they account for 95 percent of the energy in the universe today.

The two components are called *dark matter* and *dark energy*, rather unimaginative names for the two most surprising and enigmatic constituents of the universe. The nomenclature is actually confusing because it suggests that the two are related, whereas the only thing they have in common is that they do not absorb or scatter light. Otherwise, their physical properties are completely different. And their roles in this history of the universe are completely different. Dark matter dominated the past; dark energy will shape the future.

Dark matter was discovered in the 1930s by Fritz Zwicky, an astronomer at the California Institute of Technology, who was trying to understand the rapid motion of galaxies within galaxy clusters. In a cluster, the galaxies are held together by gravity and orbit about one another. Zwicky found that he could explain the high orbital speeds of the galaxies only if the clusters contain a lot more matter than is present in visible stars and gas. He proposed that most of the mass of a galaxy cluster consists of some form of nonluminous matter. Four decades later, theorists showed that individual spiral galaxies, including the Milky Way, must be embedded in a large cloud of dark matter in order for their stars to remain confined to thin disks. This prediction was subsequently confirmed by astronomical observations tracking the rapid motion of gas at the outer edge of neighboring galaxies.

In the 1980s, astronomers first began to "see" the dark matter in galaxy clusters by observing how its gravitational field bends the paths of light rays passing through it, an effect called *gravitational lensing*. To understand how one can "see" dark matter this way, just think

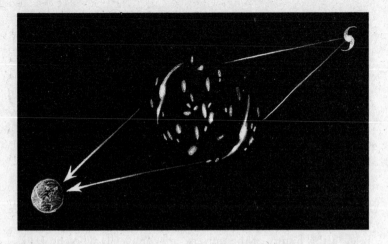

about water in a drinking glass. Water is completely transparent to light, but one can tell that it is in the glass because it bends light rays passing through it and distorts the image of whatever is behind it. Likewise, even though the dark matter in a cluster of galaxies is completely transparent, it will, because of its gravity, bend the light passing through it from a distant galaxy on its way to the Earth. Clumps of dark matter can behave as gravitational lenses, each forming a separate image of the distant galaxy behind. In a telescope image, the highly distorted, lensed images of the distant galaxies appear in the same view as the images of the nearby cluster galaxies. By analyzing the shape and light pattern of the various sources, astronomers can clearly discriminate between the two. Then, by modeling the pattern of lensed images and the bending of light, astronomers can reconstruct the distribution of dark matter in the galaxy cluster. In this way, they can effectively "see" the dark matter through its light-bending effect.

Over the last few decades, cosmologists have come to understand that the presence of dark matter was essential to the formation of all of the structures making up the universe. For a few hundred thousand years after the big bang, when the universe was still filled with hot plasma, the intense pressure of the radiation prevented the ordinary matter composed of atomic nuclei and electrons from clumping together. Because dark matter is not affected by radiation, it was free to condense into concentrated dark matter clouds, long before the ordinary matter could start to cluster. When the hot plasma cooled enough for the electrons and nuclei to combine into neutral atoms, ordinary matter was set free from the radiation. By this time, the clouds of dark matter were already in place and concentrated enough to attract the ordinary matter to their cores, where galaxies, stars, and planets then formed. If not for this assist by dark matter, galaxies would be much rarer than they are: in most of space, the dark energy would have taken over and diluted the matter away before it could ever clump. And without galaxies, there would be no stars and planets. Therefore, all of us owe our very existence to dark matter, even though its composition is not yet known.

Most physicists think that dark matter consists of an ocean of elementary particles that are electrically neutral, so that they interact very weakly with ordinary matter and do not scatter or absorb light. This would explain why the particles are not noticed even though the Earth is constantly moving through a sea of dark matter as it orbits the Sun. One way to try to detect the dark matter is to construct instruments that are sensitive enough to directly detect the dark matter particles flowing through the Earth. By placing these detectors in deep mines where they are shielded from high energy cosmic rays composed of ordinary matter, physicists hope to find a few rare events in which a dark matter particle bounces off an atomic nucleus and leaves a trail of light and ionized particles. As of 2007, many experiments are under way, but no confirmed detection has been made.

Another approach is to re-create the ultrahigh temperature

and density of the early universe in a controlled laboratory setting to see if dark matter particles can be produced in these conditions. In the near future, this will be done at giant particle colliders like the Large Hadron Collider being built at the European Organization for Nuclear Research (CERN) in Geneva, Switzerland, and the International Linear Collider, which is currently being planned.

The reign of dark and ordinary matter lasted nearly 9 billion years, during which the first atoms, molecules, stars, and galaxies formed. Over that period, dark matter accounted for over 80 percent of the energy of the universe and ordinary matter accounted for the rest. Then, about 5 billion years ago, dark energy took over as the dominant form of energy in the universe.

Although it has nearly three times the density of dark matter today, dark energy has been much harder for astronomers to detect because it does not cluster under gravity and form distinct clumps. The only way to sense its presence is to measure its repulsive gravitational effect over very large regions of the universe. The first bits of evidence for dark energy were uncovered in the 1980s and early '90s, but there were also contrary indications. The situation became clear only in the mid-1990s, when all the leading astronomical indicators converged on the same result. Ground- and balloon-based cosmic background radiation measurements, the predecessors of WMAP, showed that the light coming from the WMAP layer has not been affected by space curvature: space is flat. But it was also known by then that the sum total of all the clustered matter, both ordinary matter and dark matter, accounts for less than half the energy density required to explain the expansion rate of the universe, according to Friedmann's equation. The simplest explanation for the shortfall is that the universe is dominated by a form of energy that is transparent, so it is invisible, and smoothly distributed, so it is not counted in the census of clustered matter. To avoid clustering, this additional type of energy must be gravitationally self-repulsive, cosmologists reasoned, and therefore completely different from dark matter.

The idea that gravity can repel goes against what children are taught in school; one of the first principles of physics, dating back to Newton, is that gravity always attracts. What was less widely appreciated, until recently, is that in Einstein's theory of gravity this axiom applies only to some forms of energy, like ordinary matter and dark matter. Other forms of energy, like dark energy, *can* gravitationally repel.

The simplest and most famous example of dark energy is the cosmological constant, introduced by Einstein in 1917. You can think of it as the energy of the vacuum—that is, the energy remaining in empty space after particles and all forms of radiation have been removed. Vacuum energy is totally inert. It has the same density at every point of space and at every moment in time, no matter what is happening in the universe. When space stretches, the vacuum energy density is completely unchanged. Since the volume increases, this means that the total vacuum energy increases as the universe expands. Furthermore, the repulsive gravitational effect of vacuum energy causes space to expand even faster, creating even more space and even more vacuum energy. The result is a runaway exponential expansion and a runaway production of vacuum energy.

To understand how this runaway process got started, consider the time when galaxies were first forming and the dark energy density was much smaller than the matter density. As the universe expanded, the galaxies spread out and their density decreased, but the concentration of vacuum energy did not. Inevitably, the density of matter fell below the vacuum density. At that point the gravitational repulsion of the dark energy took hold, speeding up the expansion. Based on observations, the speedup occurred about 5 billion years ago. According to Friedmann's equation, from now on the universe will double in size every 10 billion years, placing us in an epoch of exponential growth. Gravity will continue to power this growth forever, unless the dark energy changes to another form. As described in the next chapter, such a change does indeed occur in the cyclic model.

In 1998, two groups of astronomers conducted a survey of very distant Type IA supernovae, the exploding stars described earlier, and compared their recessional speeds to those of nearby ones. The distant supernovae emitted their light many billions of years ago, so their recessional speeds could be used to determine the expansion rate of the universe in the remote past. Similarly, the recessional speeds of the nearby supernovae were used to determine the present expansion rate. By comparing the two expansion rates, these astronomers showed that the expansion is accelerating, confirming the earlier conclusion that the universe is dominated by dark energy today. Follow-up supernovae surveys, the WMAP results, and other independent checks have provided overwhelming evidence that a dark energy—dominated epoch is well under way.

The discovery of dark energy stunned scientists. Overnight, the view of the universe and of its future changed. The conventional view was that everything in the universe attracts under gravity just as ordinary matter does. For seventy years, many popular books on cosmology took this for granted and ignored the possibility of gravitational repulsion. Authors claimed that the cosmic expansion rate is slowing and that the future of the universe depends only on whether the amount of matter is sufficient to stop the expansion and cause the universe to contract to a big crunch. Now it is clear that the books were wrong. Matter ruled the universe in the past, but a new age has begun in which the influence of matter on the evolution of the universe is becoming negligible and the fate of the universe rests instead on the nature of dark energy.

If the dark energy is due to a cosmological constant, the universe will expand exponentially forever. All of the galaxies seen today will be diluted away and space will approach a nearly perfect vacuum. However, this need not be the case. Physicists have identified several alternative kinds of dark energy that might allow the universe to avoid this dismal fate. One example, called *quintessence*, is also gravitationally self-repulsive, but its density decreases slowly with time. In this case, dark energy dissipates and gives way to a new kind of evolution. A specific kind of quintessence that occurs in the cyclic model enables the universe to recover from each period of accelerated expansion and begin a new cycle.

The Big Picture

The detailed reconstruction of the last 14 billion years of cosmic history, beginning one second after the big bang, has to be counted as one of the most extraordinary human achievements. Any credible account of the origin and future of the universe must be based on what has been learned thus far.

The most basic fact is that the universe evolves. Soon after the bang, the universe was very hot and dense. But it has now expanded into a cold, dilute state. It began almost uniform and structureless and has become highly complex and elaborate. Over time, matter has been drawn together by gravity and other forces and has arranged itself in ever more complex structures: nuclei, atoms, molecules, dust, rocks, planets, stars, galaxies, galaxy clusters, and superclusters. All of this complexity arose from almost undetectable nonuniformities in the distribution of energy that existed at the one-second mark.

The matter in the universe comes in two types: dark matter and ordinary matter. Both were present in the primordial plasma emerging from the bang. Dark matter played a vital role in forming galaxy halos, and still plays a dominant role in their structure. Ordinary matter fell into the cores of the dark matter halos to from stars, supernovae, and planets. The light chemical elements—hydrogen, deuterium, helium, and lithium—were made in the hot big bang by the fusion of primordial protons and neutrons as the universe cooled. Heavier elements, including carbon, oxygen, and iron, were made by the burning of helium in stars and supernovae.

On the largest length scale that can be observed, out to 13.7 billion light-years, the structure of the universe is stunningly simple. There is no detectable curvature of space, and matter and radiation are smoothly distributed. Perhaps even more remarkable, the laws of physics seem to be the same everywhere. The expansion of space and the clustering of matter under gravity are accurately governed by Einstein's theory of general relativity. The laws of quantum mechanics, which govern the structure of atoms and molecules, the laws of nuclear and statistical physics, which govern the burning of stars, the laws of light and electromagnetism, and the laws of fluid dynamics, hold everywhere as well. The universe appears to be simple and comprehensible.

On smaller scales, gravity has worked its magic, taking the almost imperceptible nonuniformities that emerged from the bang and steadily drawing together islands of matter, dark and ordinary, which then collapsed into galaxies, stars, and planets. Gravity governs the structure of stars, heating the gas to temperatures where hydrogen can burn into helium, and helium into heavier elements. Gravity holds planets in orbit around stars. Gravity drives the collapse of stars, leading to violent supernova explosions, within which the heavier chemical elements are formed. And gravity produces the giant black holes found at the centers of most large galaxies, which swallow gas and stars and are responsible for the most violent and energetic phenomena in the universe.

The most puzzling discovery is that for the last 5 billion years, the formation of new structures in the universe has ceased and the universe has begun to become simple and uniform again. This strange turn of events is related to the fact that dark matter and ordinary matter, which are gravitationally attractive and can cluster into new structures, together account for less than a third of the total energy of the universe. The remainder is in dark energy, which is gravitationally repulsive and has begun to stretch the universe out and return it to a smooth, uniform state. At least at the moment, the tug-of-war between dark matter and dark energy appears to have been won by dark energy. Whether this situation is permanent is at the crux of the debate between the inflationary and the cyclic pictures of the universe.

chapter three

Two Tales of One Universe

It was the best of times, it was the worst of times, it was the age of wisdom, it was the age of foolishness, it was the epoch of belief, it was the epoch of incredulity, it was the season of Light, it was the season of Darkness, it was the spring of hope, it was the winter of despair.

—Charles Dickens, *A Tale of Two Cities*

For cosmologists, this is the best of times. Since the beginning of the 1990s, progress in the field has been phenomenal. One successful experiment after another has been performed that has enhanced our knowledge of the universe, making it possible to test competing views of its history. As a result, many ideas have fallen by the wayside. In 1996, at an international meeting held at Princeton to discuss the long-term future of cosmology, many different models were still in play. Three years later, at the meeting the two of us organized at the Isaac Newton Institute for Mathematical Sciences in Cambridge, only the inflationary model, with the addition of dark energy, remained viable.

When the WMAP image appeared four years later, in 2003, cosmologists worldwide breathed a collective sigh of relief that the new findings were consistent with the sole surviving model. Astrophysicist John Bahcall, giving the concluding remarks at the WMAP press conference, accurately expressed the prevailing attitude: "The most revolutionary result [obtained from the WMAP image] is that there are no revolutionary results. WMAP has confirmed with exquisite precision the crazy and unlikely scenario that astronomers and physicists cooked up based upon incomplete evidence."

Bahcall described the inflationary model as "crazy and unlikely" because the current version is a patchwork quilt sewn together from disparate ideas added over the previous two decades, plus the assumption of a particular, odd mixture of ordinary matter, dark matter, and dark energy. "Incredibly, everybody got it essentially right," he said, expressing a widely held view that since the inflationary model was the last surviving model, it must be correct.

At the time of the WMAP announcement, the cyclic model was still new and unfamiliar. Bahcall and other astrophysicists were used to comparing models that incorporated small variations on the basic inflationary model, differing in only one or two details. But the cyclic model is entirely different. It turns cosmic history upside down and introduces numerous novel and surprising elements at once. Some new elements come from fundamental physics, some from general relativity, and some from cosmology. As the cyclic model has developed and its principles have become better known, astrophysicists and physicists have begun to pay close attention. But in 2003 neither Bahcall nor most other astrophysicists were aware that the WMAP's exquisite confirmation of the inflationary predictions was simultaneously an equally exquisite confirmation of the cyclic picture.

Not only do both models fit the WMAP data with the same precision, but they also resolve the same three puzzles about the state of the universe at the start of Act Two. The first puzzle is the *homogeneity problem:* why was the universe so incredibly uniform just one second after the big bang? To appreciate this puzzle, consider the region of space visible today, extending about 14 billion light-years in all directions. When the universe was only one second old, this region was extremely uniform and a few billion times smaller than it is today, a few *light-years* across. But particles and radiation, even when moving at the speed of light, cannot travel farther in one second than a *light-second.* So one second is simply not enough time for the matter and radiation to spread out over a light-year and mix into a smooth, homogenous soup. The only way to explain the uniformity is if, somehow, the matter and radiation had been produced in an almost perfectly uniform state in the first place.

The second big puzzle is almost the opposite of the first. Even if it were somehow possible to create matter and radiation in a perfectly uniform state, that is not really what is needed. There must also be slight variations in the density from place to place at the start of Act Two, since it is these variations that clump under gravity to form galaxies and other large structures. The variations have to match those seen in the WMAP picture and all of the results of galaxy surveys. Explaining how the right kind of variations were in place at the start of Act Two is an equally challenging mystery called the *inhomogeneity problem.*

The third major puzzle is that, as the WMAP measurements confirm, space is extraordinarily flat on large scales. That is, parallel light beams neither converge, as they would if space were positively curved, or diverge, as they would if the curvature were negative. Einstein's theory of gravity allows for either possibility, but the universe has ignored them and chosen neither. The puzzle deepens when you realize that a flat universe requires a perfect balance between the expansion rate and the energy density. Any slight imbalance will grow, causing space to become increasingly curved in either the positive or the negative sense. The *flatness problem* requires an explanation for why the universe emerged from the big bang in this perfectly balanced state, a balance so good that it still holds with high precision nearly 14 billion years later.

All three puzzles—the homogeneity, inhomogeneity, and flatness puzzles—seem impossible to solve if the universe emerged from a big bang in the manner that Friedmann, Lemaître, Gamow, and others envisaged. A hot big bang of the type they conceived is a violent process with nothing to prevent wild variations in the density and curvature from place to place. So both the inflationary and the cyclic models attempt to introduce new elements to tame the big bang and to resolve these puzzles.

After that, the similarities between the two models end. The cosmological puzzles are resolved using completely different mechanisms operating at different times in cosmic history and at different energies. Even more curious are their different outlooks on cosmic history and the prospects for the future. Below, a brief overview of both cosmic tales shows how these qualitative differences arise.

The Inflationary Tale:
Heading for the Worst of Times

According to the current version of the inflationary model, the creation of new galaxies and larger-scale structures has ended. In the billions of years to come, the local group of galaxies surrounding us will remain bound together by gravity, but the hundreds of billions of galaxies lying beyond will recede at an accelerating pace. They will appear redder and dimmer as the light they emit becomes more and more stretched on its way toward the Earth. Eventually, they will disappear from view. The Milky Way and a handful of neighboring galaxies will remain clustered together but otherwise will be left all alone, surrounded by a vast, empty expanse of space. Gradually, over trillions of years, all the stars in the local group of galaxies will burn out. Over a much longer time, even the matter we're made of will slowly decay away, until nothing but dark energy is left. According to the current inflationary picture, this cold, bleak future is all that awaits us, and it is likely to last forever.

Let's now run through the inflationary story, paying special attention to how inflation is supposed to establish the conditions at the beginning of Act Two. In the inflationary picture, the big bang was the moment of creation. How exactly this happened remains unexplained. The universe is simply assumed to have appeared out of nothing, filled with all kinds of exotic matter and energy, at nearly infinite temperature and density. Cosmologists differ on the precise properties of this starting state, but many believe it would have been wildly turbulent and nonuniform, with huge variations in density and temperature from place to place, and with space curved and warped in unpredictable ways.

As we have emphasized, the existence of a beginning to time is an unproven assumption, based on using Einstein's theory of general relativity to extrapolate the expansion of the universe back in time and finding that the density and temperature reached infinite values about 14 billion years ago. Cosmologists understand that this infinity indicates a mathematical breakdown and that Einstein's theory of gravity has to be replaced by new physical laws. Nevertheless, in the inflationary picture, the presumption is made that even after the new physical laws are found and understood, this moment will turn out to be the beginning of the universe. If this idea is right, the only way to explain how the universe became so large, smooth, and flat is that the creation event was immediately followed by a spectacular burst of expansion.

According to the model, this magic is worked by introducing a special ingredient known as *inflationary energy*, which, combined with gravity, drove an astonishing amount of expansion in a fleeting interval of time. Inflationary energy can take a wide variety of forms. The duration of inflation depends on the particular choice. In a typical case, inflation lasts a mere 10^{-30} (or 0.000000000000000000000000000001) seconds, during which the universe doubles in size every 10^{-35} (or 0.00000000000000000000000000000000001) seconds. This corresponds to doubling in size 100,000 times in 10^{-30} seconds.

It is hard to appreciate what a huge amount of expansion this is. Imagine that you and a friend stand toe-to-toe separated by a single hydrogen atom, one ten billionth of a meter across. Then suppose that you double your separation over and over. After one doubling, you are separated by the width of a hydrogen molecule. After ten doublings, a virus could slip in between. After twenty-seven doublings, your toes are about a centimeter apart. After thirty-three doublings, a gap of almost a meter has opened up. After seventy-five doublings, the separation is greater than the size of the solar system; it is greater than the Milky Way after 110 doublings, and greater than ten billion light-years after 120 doublings. And this is only the start! If inflation lasts for 10^{-30} seconds, the doubling repeats at least another 99,880 times, an almost inconceivable magnification.

If space is stretched so rapidly, any curves or warps created during the big bang are ironed out, just like the wrinkles in a sheet are smoothed away as it is pulled taut. Similarly, energy is spread out uniformly across space. Matter and radiation are diluted away by the expansion, but inflationary energy is not: as the size of the inflating patch grows, the inflationary energy just becomes more smoothly

distributed while its density remains nearly constant. In this way, the inflationary model attempts to tame the wild conditions created at the big bang, producing the smooth, flat universe that is required for the beginning of Act Two.

How does inflationary energy create the exponential expansion? The key is that it is hugely dense, gravitationally repulsive, and undiluted by the expansion of the universe. The reader may feel a sense of déjà vu. This behavior, where the energy density remains fixed and the expansion of the universe accelerates, is qualitatively similar to the effect of dark energy. But there is a huge quantitative difference. The concentration of energy needed to drive inflation is a googol (10^{100}, or 1 followed one hundred zeros) times greater than the concentration of dark energy observed today. Inflationary energy dominates the universe for only a split second after the big bang but, being far more concentrated than dark energy, it causes an enormously greater rate of acceleration. Everything must happen incredibly quickly during the inflationary epoch because the smoothness of the universe must be achieved before even one second passes.

Once inflation has made the universe smooth and flat, the inflationary energy must decay into the hot plasma that is required to fill the universe at the beginning of Act Two. The decay mechanism is one of the most remarkable aspects of inflation, because it creates slight nonuniformities in the plasma of just the right type to act as seeds for galaxy formation and generates the hot spots and cold spots seen in the WMAP layer, thereby providing a possible solution to the inhomogeneity problem.

According to the inflationary model, the decay of inflationary energy into radiation occurs through a chance process governed by quantum mechanics, creating a random energy distribution of just the form needed to later seed galaxies. The quantum decay of inflationary energy is very similar to what happens when one chemical element undergoes radioactive decay into another, like plutonium into uranium. Plutonium has a half-life of twenty-four thousand years, meaning that any particular plutonium atom has a fifty-fifty chance of decaying into a uranium atom in that time. Imagine starting out with a bar of pure plutonium. After twelve thousand years, over a quarter of the atoms, chosen at random, will have turned into uranium, creating tiny pockets of uranium scattered throughout the

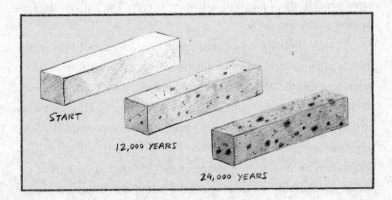

START

12,000 YEARS

24,000 YEARS

bar. After twenty-four thousand years, half the atoms will have decayed, and the pockets will merge into large islands of uranium, giving the bar a patchy appearance.

Inflationary energy decays through a similar process. In some regions of space the inflationary energy converts to radiation sooner, in other regions later. Once the energy is in the form of radiation, the density falls quickly as the universe expands, so regions that convert to radiation sooner end up cooler. Regions converting later retain a high energy density and end up hotter. The random character of the decay of inflationary energy results in a patchy universe, a bit like the plutonium bar, with hot and cold regions spanning a wide range of sizes. One of the most remarkable successes of inflation is that the variations in temperature and density from one region to the next, produced by the decay of inflationary energy, can successfully account for the hot and cold spots in the WMAP picture and for the seeds that form galaxies.

The random decay of inflationary energy was first analyzed in the early 1980s by several physicists, including Paul. Their results showed that after inflation is complete, ordinary matter, radiation, and dark matter vary over space in the same spatial pattern. This is to be expected because they are produced at the same time through the same process, the decay of inflationary energy. The pattern itself is known as *scale-invariant noise.*

A scale-invariant noise pattern appears at first glance to be completely random, but closer examination yields a subtle surprise. Consider, for example, taking a small patch of the WMAP image shown in chapter 2 and representing it as a surface on which hills and valleys correspond to the up-and-down variations in temperature. Along any line drawn on the sky, such as the leading edge shown in the illustration on page 57, the temperature hills and valleys appear to follow a randomly undulating curve. But if the pattern is scale-invariant, the curve can be represented as a sum of sinusoidal waves with similar heights for all wavelengths. This is a very striking signature that can be easily distinguished from other patterns.

The realization that the inflationary model predicts a scale-invariant noise pattern of density variations was a thrilling surprise. A decade earlier, cosmologists Edward Harrison at the University of Massachusetts, James Peebles at Princeton, and Yakov Zel'dovich of Moscow State University had independently pointed out that scale-invariant density variations are exactly what is needed to explain the pattern of observed structures in the universe, but they'd had no idea how to generate this pattern. The inflationary model provided the first plausible physical mechanism capable of performing the trick.

Nearly twenty years later, the WMAP snapshot showed with

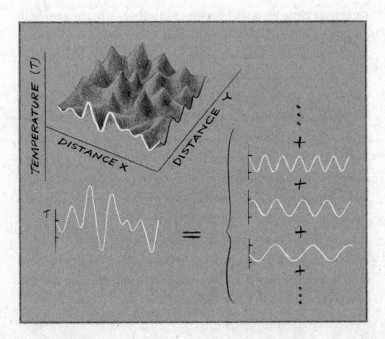

great precision that the variations in temperature and density really had the scale-invariant form that had been conjectured. So, it seemed, the inflationary model not only explains why the universe became big, smooth, and flat but also how the small imperfections needed to produce galaxies and stars were formed. No wonder cosmologists were persuaded that inflation really must have happened!

According to the inflationary model, the radiation-dominated epoch started about 10^{-30} seconds after the big bang when the inflationary energy decayed into an extremely hot plasma of elementary particles of all types. The plasma contained radiation, matter, and antimatter in roughly equal proportions. Antimatter particles are almost the mirror images of matter particles, with the same mass but the opposite charge. If matter and antimatter particles are put together, they annihilate each other and produce a burst of radiation. The reverse process, whereby radiation produces pairs of matter and antimatter particles, is also possible. In the hot early universe, both processes happened all the time, so that matter and antimatter were equally abundant. But according to unified theories of particle physics, matter and antimatter particles are not precise mirror images of each other. The tiny differences in their physical properties, led to the generation of a slight excess of matter over antimatter within the first billionth of a second after the big bang. After this time, for every 10 billion antiparticles in the plasma, there were 10 billion plus one matter particles. As the universe expanded and cooled, each antiparticle was annihilated with a particle, so that by the time the universe was one second old, only the excess matter particles remained. The surviving matter particles were in the form of electrons and quarks. Later, the quarks were joined together in threes to form protons and neutrons. After the one-second mark, neutrons and protons fused together to form atomic nuclei.

The story from then on is as described in chapter 2, consisting of the radiation epoch followed, at 75,000 years, by the matter-dominated era. At 380,000 years after inflation, the first atoms formed, as nuclei and electrons became bound to each other, and the hot plasma became a transparent gas of neutral atoms and freely streaming light. As radiation and matter ceased to interact, matter started to cluster into galaxies, stars, and planets.

There is one final sting at the end of the inflationary tale. After about 9 billion years, the tiny residue of dark energy began to dominate the universe as both matter and radiation thinned out. Since that time, the dark energy's repulsive gravity and nearly constant density have caused space to expand at an accelerating rate. From now on, the universe will double in size every 10 billion years.

As we've already pointed out, dark energy is similar to inflationary energy in that both cause the expansion to accelerate, but they differ in energy density by a googol and give vastly different doubling times (10^{-35} seconds compared to 10 billion years). A second difference is that inflationary energy is unstable and lasts only a brief moment, while the long-term fate of dark energy is uncertain. According to the simplest versions of the inflationary model, dark energy is stable and will last forever, although it is also possible that dark energy will ultimately decay and the expansion will slow down. The expansion might even stop altogether and reverse into contraction.

For all its strengths, the inflationary model rests on a significant number of unproven assumptions. The first, that the big bang was the beginning of time, immediately forces a second. To explain how the universe became so smooth and flat within the first second, the existence of a powerful new element, inflationary energy, has to be assumed. To perform its task, the inflationary energy must have specially tuned properties. Although the tuning is an unattractive feature, cosmologists accepted it because inflation seemed like the only idea that had a chance of resolving the cosmological problems of the early universe. Then, to explain the current period of cosmic acceleration, a completely new ingredient is assumed: dark energy, which has no connection to inflationary energy. Once again, there seems to be no other way to explain the astronomical observations . . . *unless the big bang was not the beginning.* In this case, one might wonder if inflationary energy is really necessary and if dark energy might be better integrated into the story. This thought is the perfect segue to our second cosmic tale.

The cyclic model was developed in the wake of two breakthroughs: one theoretical and one observational. The first was the discovery of a new unified theory of all the forces of nature, including gravity, capable of describing the universe in a more complete and consistent manner than any theory before it. The second was the discovery of dark energy, now dominating the universe and driving it toward a simple, uniform state.

We shall describe some of the ideas behind the new unified theory, called M theory or, more generally, string theory, later in this book. But the key point for the cyclic model is that M theory allows us to take seriously the possibility that the big bang was not the beginning of time. In the M theory picture of the world, the big bang can be represented as a violent transition from a low-energy density state, like today's universe, to a high-energy density state, like the hot plasma-filled universe before the start of Act Two. The cyclic model links this idea with the existence of dark energy. According to the new cosmological model, the big bang was triggered by the decay of dark energy that existed before the bang. As we shall explain, dark energy decay is a remarkable process that can smooth and flatten the universe, and create small density variations, just as effectively as the process of inflation does in the inflationary model. So the universe emerging from the bang is naturally flat and smooth. Furthermore, since it evolves into the present universe, which is itself filled with

dark energy, the process can repeat in the future. Emerging from these basic assumptions follows a new model of the universe in which the big bang repeats at regular intervals throughout cosmic history.

The Cyclic View:
Heading for the Best of Times

The cyclic tale pictures a universe in which galaxies, stars, and life have been formed over and over again long before the most recent big bang, and will be remade cycle after cycle far into the future. Cosmic evolution consists of a series of transformations, from hot to cold, from dense to dilute, and from uniform to lumpy and back again at regular intervals spanning up to a trillion years or more. Space naturally smooths and flattens itself after each cycle of galaxy formation and before the next big bang, so the model doesn't need to include a burst of inflation.

Each cycle divides into six stages. Since the cycles repeat, the description can begin with any stage. For ease in comparing the cyclic model with its inflationary counterpart, it is helpful to start at the moment when the temperature and energy density of the universe reach their highest values.

The Big Bang: Unlike the inflationary picture, the cyclic model does not include a moment when the temperature and density become infinite. Instead, the big bang is an event that can, in principle, be fully described using the laws of physics. Before the bang, space is flattened and filled with a smooth distribution of energy resulting from the decay of dark energy. At the bang, some of this energy is transformed into smoothly distributed matter and radiation at a very high temperature, high enough to evaporate ordinary matter into its constituent quarks and electrons and to produce many other exotic particles through high-energy collisions. But from *before* to *after* the bang, the fabric of space remains intact, the energy density is always finite, and time proceeds smoothly.

The Radiation-Dominated Epoch: Since the bang creates a flat, smooth radiation-dominated universe, there is no need for an intervening inflationary epoch. Below a temperature of 10^{20} degrees, there is no major difference between the radiation-dominated epoch in the cyclic model and that in the inflationary model. Just as in the inflationary case, slight differences in the properties of matter and antimatter particles lead to a tiny excess of matter over antimatter in the hot plasma. As the universe cools, antimatter particles and matter particles collide and annihilate each other, leaving only the small excess of matter particles amid a sea of radiation. A millionth of a second after the bang, the leftover quarks combine to form protons and neutrons. At around the one-second mark, they then fuse to form the nuclei of helium and other light elements.

The Matter-Dominated Epoch: Just as in the inflationary model, at 75,000 years after the bang, matter takes over as the dominant form of energy. The first atoms form 380,000 years after the bang. The universe becomes transparent. Matter draws together under the influence of gravity to form galaxies. The epoch ends after about 9 billion years.

The Dark Energy–Dominated Epoch: In the cyclic story, dark energy is the lead character. Once matter and radiation are diluted away and dark energy becomes dominant, the expansion of the universe accelerates. The concentration of galaxies, stars, dust, molecules, and atoms—everything created since the last bang—thins out dramatically and the universe approaches an empty, uniform state with few traces remaining from any previous cycles of cosmic evolution.

The Contraction Epoch: In the cyclic model, accelerated expansion does not continue forever; if it did, a cycle would never end. A key assumption in the cyclic model is that the dark energy can decay: after a period of perhaps a trillion years, the physical properties of dark energy undergo a transformation that causes the expansion to slow down and eventually halt, leading to a phase of very gentle contraction. Once one accepts that the dark energy can slowly and smoothly decay, many interesting consequences follow.

The transformation of dark energy during the course of each cycle is similar to what happens to the energy in a spring that is stretched and then released. Shortly after the big bang, the dark energy exists mostly as "potential" energy, like the energy stored in a stretched spring. Its energy density is initially very small, negligible compared to that of the matter and radiation. But whereas the densities of matter and radiation are diluted away as the universe expands, the dark energy density remains nearly constant. When dark energy eventually overtakes matter and radiation, it is still primarily in this potential energy form, whose gravitational effect is to speed up the expansion of the universe. But after a trillion years or so, the dark energy undergoes a change, similar to that in a stretched spring when its ends are released. The dark energy turns into a mixture of potential and kinetic energy. At the same time, its gravitational effect on space reverses. The expansion of the universe slows down, and eventually switches to gentle contraction. And then the dark energy acquires the properties of a gas with very high pressure, which causes it to spread itself uniformly across space. This remarkable transformation turns out to solve many of the cosmological puzzles listed above.

At the start of the contraction phase, the dark energy density is very low, equal to the value observed today. Once the contraction begins, the energy density rises rapidly and gravitational energy, the energy stored in the gravitational field, is converted into the new high-pressure form of dark energy. This form of dark energy builds up in density much faster than other forms of energy or the curvature of space. As it dominates, it ensures that the universe remains smooth and flat as the contraction continues.

Big Crunch: Finally, the contraction reaches a "big crunch." Some of the high-pressure form of dark energy is suddenly converted into hot matter and radiation, and the universe begins to expand. The crunch has turned into a bang. Because the universe was smooth and flat before the bang, it remains smooth and flat after it. Thus, the high-pressure form of dark energy that dominates the contracting phase of the cyclic model plays the same role of solving the homogeneity and flatness problems that inflationary energy does in an expanding universe.

Upon closer inspection, one finds that the high-pressure phase of the cyclic model is capable of solving the inhomogeneity problem as well. Just as in the inflationary model, the inhomogeneities which develop in the contracting phase of the cyclic model are a result of quantum fluctuations. One of the greatest surprises of the model is that the high-pressure form of dark energy can develop density variations before the big bang which are of just the required scale-invariant form. In order to make predictions for the post-bang universe, we have to follow these scale-invariant density variations from the contracting, pre-bang universe through the bang and into the ensuing expanding stage. Assuming that the transition follows the simple, predictable process described later in this book, the resulting density variations are scale-invariant and almost indistinguishable from those produced by inflation. This is remarkable since the physical conditions for producing density variations in the two models are almost exactly reversed: slow contraction with high-pressure dark energy versus rapid expansion with inflationary energy.

The agreement between the inflationary prediction and the WMAP measurements was a prime reason for the continued acceptance of inflationary hypothesis. But as it turns out, the cyclic hypothesis agrees just as well. Current measurements of the temperature pattern offer no useful way of distinguishing between the two models.

When the cyclic model's contraction epoch is over and the universe has emerged into a new, hot expanding phase, it has all the attributes it had in the big bang epoch one cycle earlier: it is very smooth and very flat but also has slight, scale-invariant density variations. As the next cycle proceeds, the behavior will repeat. Every cycle is different in fine details because the quantum jumps are random and governed by the laws of chance. However, the average properties of the universe will be the same. In particular, created anew will be galaxies, stars, and planets like Earth on which intelligent forms of life may develop.

Reflecting on the Two Tales

We have told two very different tales of the evolution of the universe. The inflationary model follows a linear path from a fiery creation into a vacuous future. The cyclic model repeats, generating bangs and crunches, expansions and contractions. Act Two fits neatly into the middle of both tales.

The more one reflects on the two perspectives, the more the current inflationary picture appears bizarre. Consider its basic attri-

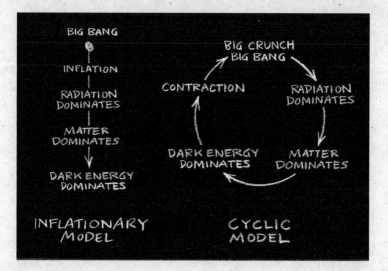

butes: For mysterious and unexplained reasons, the universe emerges from nothing into a super-dense state. A very brief period of super-rapid accelerated expansion occurs, making the universe huge and smooth. As this ends, the universe fills with radiation. Next is a brief interlude, cosmically speaking, during which galaxies, stars, and life form, after which follows an eternity of accelerating expansion. The story has a beginning but no end. The existence of life is possible only for a brief interval and then impossible for the rest of time because, for some inexplicable reason, the universe becomes forever dominated by dark energy.

The odd design of the inflationary story is perhaps a reflection of the way it was developed. Cosmologists converged on the current version by "stapling together" different ideas introduced over the course of a century: the big bang model from the 1920s, dark matter from the 1930s, inflationary theory from the 1980s, and dark energy, discovered in the 1990s. No overarching principle explains how or why any of these ideas requires the others. The big bang does not lead directly to inflation. Inflation does not require dark matter. Dark matter does not require dark energy. Each piece has been added independently and must be carefully adjusted to fit.

Successful theories can sometimes emerge from this kind of accretion process after a higher principle is found to replace the "staples" with smoother connectors. But experience shows that in many cases, a theory that must accrete more and more pieces to match the observations is headed toward failure. The classic example is Ptolemy's model of the solar system, for which epicycle after epicycle was added to try to match theory with observations. A more recent case is the steady-state model of cosmology, which was repeatedly revised as evidence for the big bang accumulated. Once the theories were in trouble, the repeated revisions were a sign that things were going wrong.

In contrast to the inflationary model, the cyclic story has an overarching principle that ties its components together: cosmic evolution is endlessly repeating with no beginning or end. The past is intimately tied to the future. Stars, galaxies, and the larger-scale structures observed in the universe today owe their existence to the period of dark energy domination in the previous cycle. And the dark energy dominating the universe today is preparing similar conditions for the cycle to come.

The role of dark energy in the two models is especially indicative of their differences. To achieve agreement with astronomical observations, the inflationary picture requires two forms of gravitationally self-repulsive energy, inflationary energy and dark energy, differing in density by a googol in magnitude. Although dark energy is the predominant form of energy in the universe today and it will determine the future, it has no connection to inflation or the rest of the story. Instead, to explain the large-scale structure of the universe, the inflationary model must introduce an entirely distinct form of gravitationally self-repulsive energy that exists for only 10^{-30} seconds and then disappears from the scene.

The cyclic model was inspired by the discovery of dark energy. Dark energy is used not only to explain the current accelerated expansion, but also to regulate the cycling. Steady cycling solves the same problems that inflation does but through different mechanisms, thereby avoiding any need for inflationary energy. The new model is also tightly interwoven. No form of energy in the past or the future ever goes out of existence. Each type reappears from cycle to cycle. When cosmologists study dark energy today, for instance, they are also studying the same type of energy that dominated during intervals in the distant past and that will dominate at intervals in the far future.

The arguments presented thus far to explain why the cyclic model is interesting and advantageous are based on common sense and simple reasoning. It is also encouraging that, since the concept was first introduced, the cyclic model has survived tests and criticisms without requiring even a single new element. Time and again, the answers have been found from within the model, based on the ingredients already present. Though this is encouraging, it is far from decisive. As matters stand now, both the inflationary and the cyclic models are worth pursuing. To determine which model, if either, will prevail in the end, we need to improve the current understanding of the fundamental laws of physics and to perform the critical experimental tests and observations that can distinguish between them.

Chapters 4 through 7 explore the connections between the two cosmological models and efforts to explain the laws governing subatomic particles and the forces through which they interact. The chapters will detour at various points to discuss some of the key developments in subatomic physics over the last century, such as quantum physics, unified field theories, and string theory, in order to explain how they transform the qualitative descriptions of the inflationary and cyclic models into serious, concrete proposals. While these chapters are perhaps the most challenging ones in the book, appreciating these ideas from subatomic physics adds great excitement to the story because they illustrate how distinguishing the inflationary and cyclic models can enhance the understanding of the universe on both the largest and the smallest scales imaginable.

chapter four

From Particles to the Cosmos

Nothing exists except atoms and empty space; everything else
is opinion.

—Democritus

Inflationary and cyclic cosmology were both inspired by attempts to
find a simple, unified description of the fundamental constituents of
the universe and how they interact. The search for a unified theory is
as old as science itself. One of the first steps along the path was taken
around 400 B.C., when the Greek philosopher Democritus proposed
that all matter is composed of indestructible, indivisible atoms. Two
millennia later, it is clear that his theory was prescient. All ordinary
matter is composed of atoms, which come in a limited number of varieties and can be organized in a periodic table according to their mass
and chemical properties. But the large number of different elements
and the overall complexity of the periodic table suggest that atoms

are not themselves the fundamental building blocks of matter. In fact, Joseph John (J. J.) Thomson's discovery of the electron in 1897 and Ernest Rutherford's discovery of the atomic nucleus in 1911 proved that the atom itself can be divided into smaller components. Ever since, the race has been on to find the basic building blocks of matter and the fundamental forces through which they interact.

At various stages during the twentieth century, physicists thought they had found the basic building blocks, only to discover they were wrong. Although electrons appear to be indivisible, physicists found that the atomic nucleus can be split into protons and neutrons. The protons and neutrons, in turn, can be subdivided into quarks, which seem to be indivisible. Quarks are never seen individually in the laboratory. They appear only in tightly bound triplets, as in the case of the proton and neutron, or paired with an antiquark (the antimatter counterpart of a quark) in unstable particles called mesons. Unstable particles last only a short time before they decay into lighter particles and radiation. By firing electrons at protons at very high energies and observing the scattering pattern, physicists have firmly established the existence of the quarks within the proton. Also, by colliding electrons with protons and protons with protons, physicists have discovered more massive quarks that are not normally seen in atoms because they rapidly decay into the lighter quarks. The collisions have also produced short-lived, indivisible, unstable particles similar to electrons, called muons and taus, and ultralight electrically neutral particles called neutrinos.

Perhaps it seems like we are beginning a never-ending list. But take heart! The list is almost complete, at least based on what is known at present. All ordinary matter that has been created in the laboratory or studied in nature can be decomposed into a combination of just the limited number of possibilities labeled "matter particles" in the table on page 72.

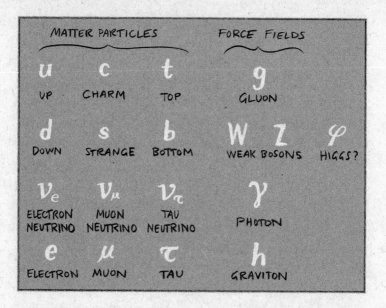

The table of matter particles somewhat oversimplifies the picture, because each of the particles comes in a number of varieties. For example, the quarks, labeled u, c, t, d, s, and b, each come in three different varieties, called "colors." A proton or neutron consists of three quarks, one of each color. Another property is called "spin": some particles act like quantum tops that rotate about an axis at only certain discrete rates, depending on the kind of particle. The quarks and the electron, muon, and tau particles come in two different spins, whereas the neutrinos come in only one. Finally, for every matter particle of each color and spin there is an antiparticle that is nearly its mirror image. Altogether, the total number of different types of matter particles and antiparticles is ninety.

Identifying the elemental constituents of matter is only half the job. Unified theories must also account for all the forces and interactions between the matter particles. Four fundamental forces have so far been identified: gravity, electromagnetism, the "strong" force, and the "weak" force. As will be explained shortly, every force is associated with a force-carrier particle: electric and magnetic forces with the "photon," the strong force with the "gluon," the weak force with the "W and Z bosons," and the force of gravity with the "graviton." These force-carrier particles, along with the Higgs particle to be discussed below, are shown on the right side of the table.

Gravity and electromagnetism are familiar in everyday life because they have effects that are commonly experienced. By contrast, the strong and weak forces are measurable only on subatomic length scales. Their effects can be studied only by probing a nucleus or colliding particles in an accelerator. The "strong" force holds quarks together inside the proton and neutron and causes protons and neutrons to stick together in the atomic nucleus. The "weak" force is an interaction between quarks, electrons, and neutrinos that is important in many nuclear fission and fusion reactions, including those that produce the energy that makes the Sun shine.

Each type of force is transmitted through a force field, which has a strength and a direction at each point in space. The strength determines the magnitude of the force exerted, and the direction determines which way the force pushes. Children experience force fields when playing with magnets. They discover that magnets can exert forces on one another without ever touching and that the force depends on the distance and direction from the magnetic poles. Later, they learn that this is so because each magnet creates a field everywhere in space and that other magnets move in response to the field. The magnetic field is invisible, but by putting a piece of paper on top of the magnet and sprinkling iron filings on the paper, they are able to visualize the field. An analogous situation holds for the four fundamental forces. The matter particles replace the magnets as the sources of the force fields. And the different matter particles in the table are distinguished according to the types and strengths of force fields they create.

The gravitational force field, whose strength depends on the mass of the object that creates it, is a special case because it has an alternative interpretation as a curving or warping of space-time. So when one throws a ball and watches it travel through the air on its parabolic path, there are two equivalent descriptions. One can say that the Earth creates a gravitational field that bends the path of the ball or, equivalently, that the Earth curves space-time and the ball's path is distorted by this curvature. This chapter focuses on the force-field picture since it relates more closely to the description of the other forces.

The Quantum Universe

Unified theories are based on subatomic particles, so the principles of quantum physics governing how matter and energy behave on microscopic scales are an essential part of the story. Quantum physics also plays a crucial role in the two theories of the early universe described in this book. Two quantum principles are especially important.

The first quantum principle is that almost all of the parameters we normally use to describe the world, like the position, velocity, or energy of an object, are randomly fluctuating and inherently indeterminate. In most cases, the randomness is significant only on subatomic scales. For example, one cannot say exactly where an electron is as it orbits an atomic nucleus. The best one can do, according to the laws of quantum physics, is compare the probabilities of the electron being at various possible locations around the nucleus. When you hold an everyday object like a cup in your hands, the positions of the subatomic particles from which it is made cannot be precisely pre-dicted. However, because there are so many particles, the random-ness in their locations averages out and the position of the cup can be predicted with very high accuracy, which is certainly a good thing if you are trying to drink from it! Even for a large object like a cup, however, Heisenberg's uncertainty principle states that it is impossi-ble to know an object's precise position and its velocity at the same time. If at some particular time you determined the position of the cup exactly, the cup's velocity would be completely uncertain. Since the velocity determines the motion of the cup, its position would be completely uncertain just one instant later—the cup could be any-where at all.

This bizarre, nondeterministic behavior of quantum mechanics means that, in general, one cannot predict the precise outcome of any physical process. All one can predict are statistical quantities such as average outcomes and probabilities for deviations from the average outcome. We call these deviations *quantum fluctuations*. In both the cyclic and inflationary models they are the seeds from which all of the structure in the universe originates: galaxy clusters, galaxies, stars, and planets, as well as the temperature variations in the cosmic back-ground radiation.

The second key quantum principle is that all matter particles and force fields come in discrete, indivisible energy packets that can exhibit both particle-like and wavelike characteristics, depending on the circumstances. Consider, for example, the electromagnetic field. As James Clerk Maxwell showed in 1865, it is possible to disturb an electromagnetic field and create waves that travel through space. Along any wave, the electric and magnetic fields vibrate at right angles both to each other and to the direction in which the wave is traveling.

In one of the most wonderful moments in the history of

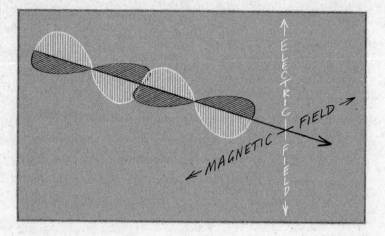

physics, Maxwell computed the speed of the wave and found it to be exactly the speed of light, which led him to the astonishing conclusion that light is nothing but an electromagnetic wave. In subsequent years, physicists discovered that visible light is only one type of electromagnetic wave. There is a broader spectrum of invisible waves, including radio waves, microwaves, X-rays, and gamma rays.

When their intensity is high, electromagnetic waves travel, combine, and interfere with one another just like the waves on the surface of the ocean. But when the intensity is very dim, a dramatic change occurs: the electromagnetic waves act like a collection of individual particle-like packets of energy called *photons*. If the light is made still dimmer, there might be only a single photon traveling alone.

The concept of a photon, or *quantum* of light, was introduced by the German physicist Max Planck in 1900, in trying to resolve a contradiction between Maxwell's theory of light and the classical theory of heat, which had been developed in the first half of the nineteenth century. The problem was that, when the two theories were combined, they predicted that a hot body would instantaneously radiate all of its heat into light waves of arbitrarily high frequency. Planck

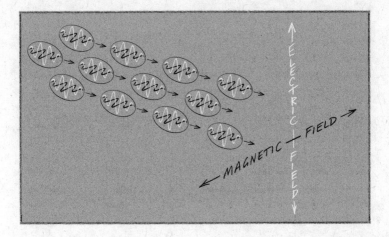

found he could only resolve this blatant conflict with reality by introducing the radical assumption that light is not a continuous wave, as Maxwell had proposed, but instead comes in discrete energy packets, called quanta or photons, whose energy is proportional to the frequency of the light. Planck showed that quantizing light in this way has the effect of shutting off the production of light at very high frequencies, so that the theory gives a sensible, finite result for the radiation emitted by hot bodies. He found that this quantum hypothesis also explained the detailed spectrum of colors emitted by hot objects, such as a red-hot iron bar or a star. Since the wavelength of light decreases as its frequency grows, Planck's law can also be stated as saying the energy per photon grows as the wavelength of light decreases.

In 1905, Einstein picked up on Planck's bold idea, realizing that the most direct way to test the quantum hypothesis is to study light sources that have a very short wavelength and are very dim, so that they emit only a few photons at a time. He considered the *photoelectric effect*, a well-known phenomenon in which electrons are ejected when light is shone on a metal surface. If Planck was right, he surmised, then shining dim ultraviolet (short wavelength) light should be very effective in ejecting electrons because, even if the photons were few in number, each photon has enough energy to smack into an electron and eject it from the metal. The short wavelength photon behaves as if it were a particle, and its impact on the electron is similar to what happens when one billiard ball strikes another. However, red (long wavelength) light is ineffective, even if it is bright, because each photon carries too little energy to eject an electron. Einstein's prediction was verified and, later, both he and Planck won the Nobel Prize for laying the foundations for quantum physics. A century later, an effect similar to the photoelectric effect is used in solar cells to convert light from the Sun into electrical energy, and quantum physics is routinely employed by physicists and engineers to develop new technologies. In fact, almost all the key technological advances over the last one hundred years—from computers to communications to chemistry and new materials—have sprung from the use of quantum physics.

Just as electromagnetic fields are composed of individual quanta called photons, the strong force field is composed of gluons, which hold quarks together; the weak force field is composed of W and Z bosons; and the gravitational field is composed of gravitons. The quanta of each type of force field have been included in the table of particles on page 72, thereby representing all of the different forces and interactions that have been observed to date.

Quantum physics, in fact, introduces a degree of unification by blurring the distinction between matter particles and force fields. Just as an electromagnetic field can act like a continuous wave or like a collection of particle-like photons, depending on the situation, so it is with electrons, quarks, and neutrinos. An electron, for example, acts like a particle when fired at the back surface of a television screen, where it travels like a bullet along a precise path and makes a pinpoint spot. But fire electrons down the open channels in a crystalline arrangement of atoms and the pattern that emerges is as if each electron has spread out into a wave that travels down all the channels at once.

The blurring of particle and wave properties led to the introduction of the concept of *quantum fields* to describe both matter particles and force fields. A quantum field describes both individual quanta that act like particles and large collections of quanta that have the sinusoidal characteristics of a continuous wave. The quantum field description allows physicists to express the fundamental laws of physics in terms of a few rules for how the different quantum fields interact with one another. This approach was considered the most promising and was intensively developed by theoretical physicists in the 1970s.

Stunning Simplicity at the Big Bang

By the 1970s, cosmologists and particle physicists felt they had the theoretical tools needed to explore what happened during the first instants after the big bang. They knew that at very high temperatures, matter would be boiled into a gas consisting of all of the matter and force-field quanta listed in the table on page 72. The quantum nature of the constituents, and their behavior under these extreme conditions, could be accurately predicted using quantum field theory.

What made cosmologists and particle physicists especially excited was the bold, new idea of *grand unification,* which suggested that a far greater simplicity might underlie the table of matter particles and fundamental forces. The central idea behind grand unification is that the strong, electromagnetic, and weak force fields, which appear to have very different strengths and characteristics at low temperatures, are actually equivalent parts of a single, unified force. The equivalence becomes apparent only at very high temperatures, such as

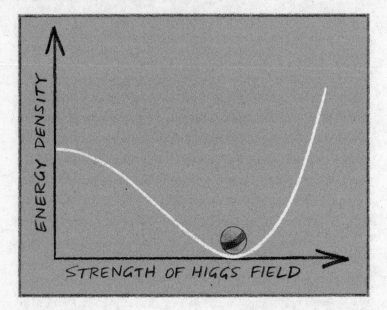

those attained just after the big bang. Since the matter particles are distinguished according to which forces they exert, they too become equivalent when the forces unify. So in the extreme conditions of the very early universe, the list of distinct forces and particles undergoes a remarkable reduction.

The device that controls when force fields and matter particles are equivalent and when they behave differently is called the *Higgs mechanism,* named after Peter Higgs of the University of Edinburgh, who first introduced the concept into particle physics in 1964. The Higgs mechanism relies on a new set of fields, called Higgs fields, which take a value at every point in space. Their role is to break the symmetry between different types of particles and forces. In a grand unified theory, different groups of Higgs fields become important at different times in the history of the early universe. To understand how Higgs fields work and their possible role in inflationary cosmology, though, it suffices to imagine that there is just one Higgs field, as we shall do for the remainder of this chapter.

The Higgs field acts like a variable light switch that controls whether forces and particles behave differently or not. If the value of the Higgs field is zero, the field is "switched off" and has no effect. Consequently, the strong, electromagnetic, and weak forces are equivalent and the matter particles behave indistinguishably from one another. This is the situation of greatest symmetry. However, if the Higgs field is "switched on," the forces split into different types and all the matter particles develop different masses, charges, and interactions. The differences depend on the *strength* of the Higgs field, which can take any positive value. A greater strength produces a greater difference between the types of particles and interactions. In this way, the Higgs field is responsible for *breaking the symmetry* between the elementary particles, leading to the complex pattern of particles depicted in the table on page 72.

The strength of the Higgs field is set by a combination of two factors, its energy curve and the temperature. The energy curve depicts how the amount of energy stored in the Higgs field depends on its strength. Its precise shape is not yet known. Different possibilities will be discussed in this chapter as we consider various cosmological models. In each graph, the potential energy density (the energy stored in the Higgs field per cubic meter of space) is shown along the vertical axis and the Higgs field strength is shown along the horizontal axis. For example, in the curve on page 80, when the Higgs field strength is zero and the field is switched off, there is positive stored energy. As the Higgs field *strength increases* (switches on), the *stored energy decreases* until it reaches zero at some positive value of the field strength. The stored energy density is now at its minimum value. So, in this example, as in grand unified theories generally, it is energetically favorable for the Higgs field to be switched on and for the symmetries between particles and forces to be broken.

Suppose the Higgs field strength is initially switched off, though. This corresponds to beginning on top of the plateau at the leftmost side of the diagram. Then, the energy curve affects the Higgs field strength in the same way that a hilly surface affects the position of a ball placed upon it. Just as a ball rolls downhill, the Higgs field rolls down the energy curve, its strength increasing until it settles at the minimum of the energy curve. Now the Higgs field is fully switched on.

The second effect determining the strength of the Higgs field is the temperature. If the Higgs field interacts with hot plasma, for example, energy from the plasma is transferred to the Higgs field, causing its stored energy density to rise and its strength to decrease. With increasing temperature, the strength of the Higgs field moves gradually from right to left (and up the energy curve) until the strength reaches zero. Then the Higgs is switched off and it has no effect on particles and forces.

The implications are profound. All of the complexity in nature—from galaxies to rocks to atoms—relies on the fact that particles such as electrons and quarks have distinct properties. Yet according to grand unified theories, the diversity is a chimera. The complexity can be eliminated by raising the temperature so that the Higgs fields are switched off. Then nature's true simplicity and symmetry emerge. The process is similar to what happens with snowflakes, which appear in an endless variety of shapes when frozen but become indistinguishable drops of water when they melt. By raising their temperature, one discovers that they are all composed of the same simple constituent.

The temperature required to switch off the Higgs fields and re-

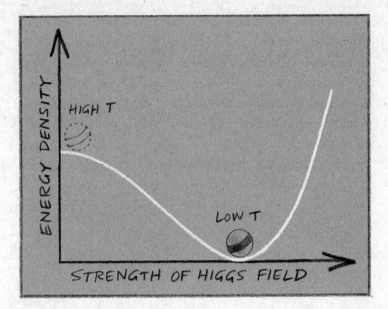

veal the underlying simplicity is extraordinarily high, about 10^{27} degrees. This is more than a trillion times beyond the range of any existing laboratory. But these temperatures were reached within the first 10^{-35} seconds after the big bang. So the idea of grand unification enticed particle physicists and cosmologists into exploring the very early universe. The siren call was this: the laws of physics get *simpler* as one goes back to the beginning; you *can* understand what happened at the big bang.

In the 1960s and '70s, most particle physicists judged cosmology to be too speculative and recommended that their students steer clear of it. But a few celebrated theorists, like Andrei Sakharov, the Russian physicist and human rights activist, and Steven Weinberg, one of the chief architects of unified field theories, were notable exceptions. Each of them had set an important example for young theorists by exploring how the natural interactions among elementary particles in the early universe could account for the fact that the universe consists almost entirely of matter today, with only trace amounts of antimatter. The mechanism they identified is, in fact, incorporated into both the inflationary and the cyclic models. As important as the research itself was the impression it left on the younger generations of physicists. The fact that world-renowned scientists would consider this kind of problem worthy of their attention sent the message that cosmology was ripe for exploration by particle physicists. By the early 1980s, a growing band of young particle theorists had begun to follow their pioneering trail and explore other puzzles lurking in the early universe. The two of us were part of this new generation.

chapter five

Inflation and the Tale of Two Cosmologists

A new idea is delicate. It can be killed by a sneer or a yawn; it can
be stabbed to death by a quip and worried to death by a frown
on the right man's brow.

—Ovid

A person with a new idea is a crank until the idea succeeds.

—Mark Twain

Both of us were educated to become particle physicists, pursuing the
dream of a unified theory. We were each drawn to the subject during
our college days when we learned that studying the interactions between the most microscopic constituents of matter had proven, up
to that time, to be the most promising way of revealing the fundamental laws that govern the universe. Just as we reached the stage of
choosing our research directions, a new discipline emerged that at-

tempted to fuse the study of elementary particles with the study of the cosmos, offering a powerful new approach for unlocking nature's secrets. The new discipline became known as particle cosmology. Its foremost triumph, inflationary cosmology, coincided with our coming of age scientifically. Our personal stories of how we became cosmologists are intertwined with the story of how inflationary cosmology was born. These narratives capture a turning point in the history of our science and are representative of the generation of optimistic young scientists who poured into this exciting field during the early 1980s.

Paul's Story

My introduction to cosmology came about two years before Neil's, and three years before we first met. I was in my second year as a junior fellow in the Society of Fellows at Harvard University, after having spent four years as a graduate student obtaining my Ph.D. degree in elementary particle physics. One day I received a slip of paper in my mailbox imprinted with this message:

Joint Theoretical Seminar
Wednesday, March 5, 1980
Inflationary Universe
Dr. Alan Guth, SLAC
Jefferson 250
Harvard University
4:30 P.M.

Every week during my nearly six years at Harvard, I'd received a similar invitation to the Boston area's most important physics seminar.

Everyone, from first-year graduate students to the most prestigious faculty, made it a weekly ritual to attend. It was *the* place to hear the latest ideas in fundamental physics.

A talk on cosmology was unusual, though. Generally, the seminar focused on elementary particle physics. Most of the senior faculty in attendance, including Nobelists Steven Weinberg and Sheldon Glashow; my Ph.D. thesis adviser, Sidney Coleman; Howard Georgi; and their MIT counterparts, as well as the majority of the audience, worked on quantum fields and elementary particle physics.

I had never taken a course on cosmology or studied it on my own. I had never heard of the young postdoctoral fellow giving the talk, and there was no paper on the subject that I could consult. (Guth's first paper on inflationary cosmology would not appear for several months; I later learned that he was in the midst of a long lecture tour when he gave this talk, and that many of the senior faculty were aware of what he had accomplished. But most of the audience was in the same position that I was, not knowing what the title meant.) So when I entered the seminar room, I frankly had low expectations. Little did I know that I was about to experience both the most inspiring and the most depressing talk I would ever hear.

After the perfunctory introduction, Guth rose from his seat, arranged his plastic transparencies, switched on the overhead projector, and began to speak. The talk was masterful from the very start. The first third was a great relief. Instead of launching into his new idea right away, the way most speakers would, Guth used the first twenty minutes to deliver a lightning review of the fundamentals of cosmology in a concise language that a particle physicist could immediately absorb. When Guth's succinct review came to a close, I realized that the big bang picture was surprisingly simple: it was based on a few assumptions, some basic physics, and a small number of simple equations. It was not the complex and intimidating subject I had anticipated. Yet the result was a rich and powerfully predictive theory supported by compelling evidence. This education alone made coming to the talk worthwhile, I thought to myself, while Guth paused to answer a few questions before beginning the second third of his talk.

I could not have been more surprised at what Guth said next. Having spent the first twenty minutes building up the big bang picture, Guth now proceeded to point out its flaws: the homogeneity problem, the flatness problem, and the overproduction of very massive particles called magnetic monopoles.

Magnetic monopoles are hypothetical particles that have only one pole of magnetic field (either north or south) instead of the usual two. None has ever been discovered in nature, but according to grand unified theories, they exist and are very massive, more than a quadrillion times the mass of a proton, or as massive as a grain of sand. Furthermore, so many should have been produced when the Higgs fields switched on that they would have completely overwhelmed all other types of matter and radiation, upsetting all the successful predictions of the big bang model. I was somewhat aware of the monopole overproduction problem because a fellow graduate student at Harvard, John Preskill, had pointed it out a few years earlier. But I had never heard of the other flaws, and they seemed much more serious. How could a theory work so well and, at the same time, so poorly?

Finally, with twenty minutes remaining, Guth came to the subject promised by the title of his talk, the inflationary universe. He then presented a remarkable idea that could resolve all three problems with one stroke. He began by turning the audience's attention to the Higgs field and its energy curve. He reminded us that as the universe cools, the Higgs field strength is initially zero. Then, the Higgs field switches on, moving from the plateau on the left side of the figure to the minimum of the curve as its strength increases. This

change in Higgs field strength is the important feature for particle physics because it causes the unified forces and matter particles to split into different types.

But then Guth asked us to turn our attention to the simultaneous motion in the up-down direction, the change in the contribution the Higgs field makes to the vacuum energy of the universe. Although particle physicists had been ignoring this feature, it might have a profound effect on cosmology, Guth emphasized. To illustrate his idea, he focused specifically on the "grand unified" Higgs field responsible for distinguishing the strong force from the other forces. This Higgs field is supposed to switch on at a temperature of about 10^{27} degrees, about 10^{-35} seconds after the big bang. But the energy curve does not have to look like the example discussed above. Different shapes are possible, and Guth asked the audience to consider one that looks like this:

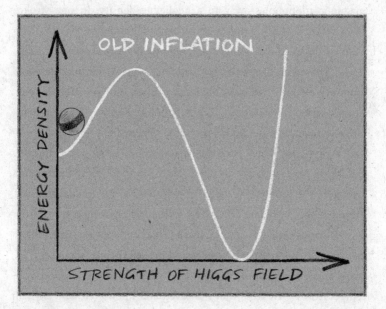

OLD INFLATION

ENERGY DENSITY

STRENGTH OF HIGGS FIELD

In this case, Guth explained, the universe starts the same way as before. At the searing temperatures near the big bang, the Higgs field is switched off and its value is zero, as indicated by the ball on the left side of the figure. But there is an energy barrier in this case that prevents the Higgs field from simply rolling downhill as the universe expands and the temperature decreases. So as the universe cools and expands, the Higgs field remains trapped at zero strength by the barrier. After 10^{-35} seconds, all the energy associated with matter, radiation, and monopoles is diluted away and the only energy that remains is the constant positive vacuum density due to the Higgs field. This energy density corresponds to the height of the ball in the diagram.

This constant energy density acts just like a cosmological constant. Its gravitational field is repulsive, causing space to expand exponentially. In fact, the energy density in the Higgs field is so high that the universe doubles in size every 10^{-35} seconds. The extraordinarily rapid expansion smooths and flattens the universe and dilutes the magnetic monopoles away to a negligible density. In one stroke, three of the biggest problems of the big bang model are solved!

Guth called this period of exponentially rapid expansion "inflation." (The name, I learned later, had been suggested by my thesis adviser when Guth had presented the idea to him privately several months earlier.) In 10^{-30} seconds of inflation, the universe would double in size 100,000 times, much more than needed to solve the cosmological problems. Then, once the Higgs field settled down at the true minimum of the energy curve and gave up its energy into matter and radiation, the universe would be in just the smooth and flat state assumed at the beginning of Act Two in the hot big bang picture.

I was stupefied. Guth had pieced together concepts from three disparate disciplines of physics—grand unified theories, general relativity, and thermodynamics (the study of how systems change with temperature)—all areas that I enjoyed and knew well, and he had

applied them to a subject I knew nothing about, cosmology, with revolutionary effect.

Then came the crash. The most exhilarating talk that I had ever heard changed direction in a matter of moments. The talk had taken nearly sixty minutes already. Although there was no strict time limit, by convention Guth had only a few minutes left for final comments. In those few minutes, he explained why the bold and beautiful inflationary idea was doomed to dismal failure. The very mechanism that solved the cosmological problems made it impossible for the rapid expansion to end. Inflation, once begun, would continue forever.

Energy curves analogous to those that Guth was considering are used to describe many everyday phenomena, such as the transformation from water to ice. Above zero degrees Celsius (or thirty-two degrees Fahrenheit), water molecules are sufficiently randomized that they form a liquid, analogous to the state where the Higgs field is zero. As the temperature falls below zero, it is energetically favorable for the molecules to organize themselves into an orderly crystalline arrangement, ice. However, to get into that arrangement, the molecules must move through a series of higher-energy configurations that are disfavored. This means that there is an energy barrier, like the one in Guth's inflationary model, that must be traversed before crystals of ice can form. Any impurities in the water or scratches in its container will help form seed ice crystals, which grow and cause the water to freeze. But if the water is pure and the container is smooth, the energy barrier will enable the water to be cooled to temperatures well below zero and still remain a liquid. This phenomenon is known as *supercooling,* and it occurs in many different kinds of phase transformations that physicists study in the laboratory. Even under ideal conditions, supercooling does not last forever; the water molecules continue to jiggle around and eventually, by chance, form a grain of ice somewhere in the liquid, which acts as a seed crystallite. Molecules in the water rapidly stick to it, so that the crystallite grows rapidly, converting liquid to solid as it goes. In a large tank of supercooled water, many such grains form, grow, and eventually merge to complete the transformation from water to ice.

My thesis adviser, Sidney Coleman, had written the seminal papers on the analogous process for the Higgs field, so I was familiar with this idea when Guth began to describe it. Coleman had introduced the terms *false vacuum,* to describe the high-energy phase in which the Higgs field is switched off (the left side of the energy curve), and *true vacuum,* to describe the lowest point on the energy curve where the Higgs field lies after it has been switched on. Then Coleman had described how quantum fluctuations can cause the Higgs field to jump over the energy barrier at some random point in space, creating a small bubble of true vacuum in which the Higgs field is switched on (the light gray bubbles in the figure on page 93), surrounded by the rest of space in which the Higgs field is switched off (the darker region). The bubble grows at the speed of light, switching on the Higgs field and converting false vacuum to true vacuum as it goes. Other bubbles form, and eventually they all coalesce to complete the transformation from false to true vacuum. The high-energy collisions between the bubble walls convert their energy into hot matter and radiation. And so the transformation can be completed.

Or can it? In the last few sentences of his lecture, Guth explained why inflation itself prevents the completion of the phase transition: the space between the bubbles inflates so rapidly that the bubbles are never able to traverse the interval that separates them. This may seem surprising, since the bubbles are growing at the speed of light and Einstein's special theory of relativity guarantees that nothing travels faster through space than light. The subtlety is that his general theory of relativity places no restriction on how fast space

can stretch. It is possible for space to stretch so fast that light cannot keep up. In the case of inflation, the space in between the bubbles stretches so fast that the bubbles never collide even though they are growing outward at the speed of light. The result, Guth and his collaborator Erick Weinberg had shown, is that the bubbles remain empty and the space between them remains empty. The result is a "cold, Swiss cheese" universe that nowhere looks like the universe observed today. The inflationary expansion that had seemed a magical elixir a few minutes before had suddenly turned into a toxic poison. And there the talk ended.

I simply could not believe that such a beautiful idea could fail so catastrophically. Immediately after the lecture, I tried to see if I could find a flaw by examining each step of the analysis and by checking it using other methods. As I became more frustrated, I became more

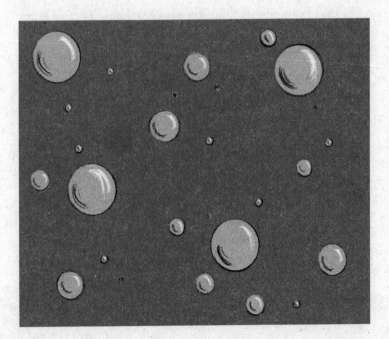

excited. I realized that whether inflation could be saved or not, there were important problems to be solved. If inflation failed, it only meant that a new solution to the cosmological problems had to be found. I figured that I would divert my attention from my ongoing research for a month or two to see if I could concoct a solution; then, after this brief sojourn into cosmology, I would return to my research on quantum field theory. (Needless to say, I was naive: twenty-six years later, I am still working on cosmology.)

By the fall of 1980, I had learned a lot more about cosmology and was beginning to develop several new ideas of my own, getting wonderful advice and support from Guth, who had moved to Boston to accept a junior faculty position at MIT. We soon became lifelong friends. I also had begun reading about bubble nucleation and phase transformations in liquids and solids, hoping to find an idea to save the inflationary model.

The next summer I hit upon a possible solution, inspired by my reading about an unusual kind of phase transformation with the strange name "spinodal decomposition." This transformation occurs in mixtures of different types of liquid helium. In this case, the energy curve has a barrier, as in Guth's inflation model, but its height shrinks to zero as the temperature falls, transforming the energy mountain into a very flat energy plateau. Then there is no need for bubble nucleation to complete the transformation. With no barrier holding it back, the helium slowly and smoothly relaxes to its low-energy state.

I realized that an analogous situation might occur for the Higgs field. Then there could be plenty of inflation while the Higgs field is high on the energy curve; yet, with no barrier to hold it back, the field would eventually evolve down the flat plateau to its low-energy state and inflation would end. An example of spinodal decomposition occurs, I realized, in a Higgs field model studied several years earlier by Erick Weinberg and Sidney Coleman, who had pursued it for a technical purpose having nothing to do with bubble nucleation or cosmology. I felt a spark of hope that this new kind of phase transformation could save inflation by enabling it to end in a smooth and continuous way.

In August 1981 my wife, Nancy, and I moved with our four-month-old baby, Charlie, to Wayne, Pennsylvania, about twenty miles outside of Philadelphia; I was beginning a junior faculty position at the University of Pennsylvania, and Nancy was starting her teaching career in the Art History Department at Bryn Mawr. Within days of arriving at the new house, I took off for Banff, Canada, to participate in a summer workshop on particle physics. I decided to try slipping in my idea about spinodal decomposition at the end of my talk, to gauge the audience reaction. I was still inexperienced at giving talks, though. Having originally committed to speaking on a different topic, I felt obligated to cover that subject first. But because this first part went overtime, the introduction to the new spinodal decomposition idea during the last instants of my presentation was too rushed for anyone to appreciate. Nevertheless, preparing the talk had led me to consider a problem with the approach: the Coleman-Weinberg model achieves its peculiar energy curve with its very flat plateau only at the expense of a highly artificial adjustment of the parameters that determine how the Higgs field interacts with itself and other forms of matter. Such an adjustment is called *fine-tuning*. If the inflationary model required fine-tuning, this seemed to me a serious flaw, because the whole point of inflation was to avoid having to assume finely tuned initial conditions when the universe emerged from the big bang. Trading that fine-tuning problem for a different fine-tuning problem in the inflationary model did not seem to me like much progress.

Upon returning from Banff, I decided that the first thing to do was determine whether the fine-tuning could be avoided. On my very first day on campus, a bright young graduate student, Andreas

(Andy) Albrecht, appeared at my office door to ask if he could work with me. I suggested that he try to see if the spinodal decomposition idea really worked in detail and if the fine-tuning could be avoided. Andy had been expecting a project in particle physics and, just like me a few years earlier, he knew little about cosmology. But he learned quickly and was soon hard at work. He developed a series of computer programs that enabled us to study what happens when the degree of fine-tuning is reduced and the energy plateau is not flat. We also examined gravitational effects on spinodal decomposition, hoping that they might help avoid the fine-tuning, though our results kept showing otherwise.

We did not know it, but thousands of miles away, in Moscow's Lebedev Physical Institute, a young theoretical physicist named Andrei Linde was following a similar path. Linde, too, had been enthralled and yet disappointed by Guth's inflationary theory. He, too, had become fixated on saving the idea. And he, too, was thinking about a slow, continuous phase transition to save inflation, though he had not made the connection with spinodal decomposition and was unconcerned about the issues that were preoccupying us. For example, he did not notice the crucial gravitational effect that reduces the kinetic energy of the Higgs field, slowing its motion and greatly enhancing the amount of inflation. In almost all inflationary models considered ever since, this gravitational effect is essential for obtaining sufficient inflation. Furthermore, Linde seemed less concerned about the fine-tuning issue than we were, perhaps figuring that the advantages for cosmology outweighed the disadvantage of fine-tuning. So he proceeded to distribute a preprint without really addressing the fine-tuning issue. When we subsequently published our independent results, we showed that the inflationary model can work, in principle, but we tempered our conclusion with a presentation of our extensive studies demonstrating that neither the gravitational effects nor any other of our other ideas could resolve the fine-tuning problem. As it turns out, no one has solved the fine-tuning problem thus far, and so it remains one of the worrisome aspects of the inflationary model, even as inflation has become a dominant theoretical idea in cosmology.

In both Linde's and our solution, the energy barrier for the Higgs field disappears just after the universe begins to supercool. As a result, the energy curve ends up with a very flat plateau separating the false vacuum (Higgs field equal to zero) from the true vacuum.

The Higgs field rolls beginning with near-zero strength (where the ball is shown) and heading toward the lowest point on the energy curve. With the barrier gone, the Higgs field strength grows continuously (moving toward the right in the figure), and the energy density falls smoothly. Gravity, as Andy and I had discovered, acts like a frictional drag force that slows the field down. As a result, the Higgs field

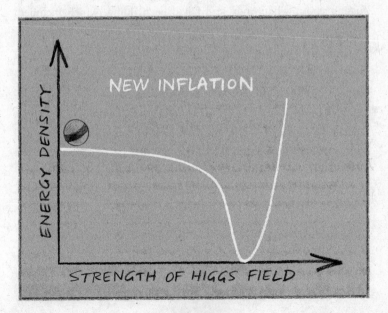

spends a very long time near the top of the hill where the energy density is nearly constant and positive. During this period, the universe inflates, just as it did in Guth's model. But the inflation only lasts a finite time until the Higgs field reaches the edge of the plateau and falls off the precipice.

Suddenly the inflationary idea was alive again, although not exactly as Guth had envisioned it. Linde called the improved approach "new inflation," a name that has stuck.

As Andy and I were preparing our paper, Michael Turner, a professor of astrophysics at the University of Chicago, visited our group at Penn. He and I instantly resonated with each other. As with myself and Guth, Turner and I and our families have remained close friends. Turner had recently received Linde's preprint and had begun talking with Frank Wilczek, then at the Institute for Theoretical Physics at Santa Barbara, about how the universe might reheat to a high temperature at the end of inflation. Together, we had an exciting discussion about what Andy and I had been doing, which included developing some computer programs that might be adapted to studying the reheating of the universe. We immediately agreed to work together on solving the reheating problem, and thus began a long series of fruitful collaborations between the two of us. Turner has a thorough knowledge of astrophysics and cosmology, as well as particle physics, and he has an infectious enthusiasm, so working together was also a pleasurable experience.

As we were completing our first project, Turner and I began to discuss a shared deep-seated worry: the possibility that new inflation might be too successful. It seemed that inflation makes the universe too smooth, so that there are none of the nonuniformities needed to form galaxies. Only one hope remained for creating the nonuniformities: quantum fluctuations.

Quantum theory says that all physical quantities fluctuate. The energy, location, and velocity of subatomic particles are some examples. Another example is the value of the Higgs field, and the associated energy density. During inflation, as at all times, the energy fluctuates on microscopic scales so that, at any instant, the distribution of energy is never perfectly smooth. Normally, one could ignore quantum fluctuations when discussing cosmology because the random fluctuations on microscopic scales average to zero when describing properties on length scales of cosmological interest. With inflation, though, the story is different. A slightly wrinkled distribution of energy on microscopic scales is almost instantaneously stretched by inflation until it spans cosmic distances. So when inflation is over, the distribution of energy is not perfectly smooth after all; rather, it is imprinted with the quantum fluctuations that were produced during inflation and stretched until they span volumes of astronomical sizes.

Turner and I realized that the quantum fluctuations could spell triumph or tragedy for inflation. If the quantum fluctuations had the right properties, inflation could explain the large-scale structures in the universe, chalking up another victory. If the quantum fluctuations caused the distribution of energy after inflation to be too bumpy, the inflationary idea was dead. Consequently, all the work on inflation up to this point now rested on the outcome of the calculation of quantum fluctuations.

The problem of tracking energy density fluctuations in general relativity was, at the time, famously difficult. In principle, one is trying to find how the fluctuations vary in space and time, but measurements of space and time depend on the observer, according to relativity. To be sure of the answer, a method is needed that follows the evolution of the density fluctuations in a way that does not depend on the choice of the observer. As we began to investigate the issue, Turner suggested that we use a sophisticated mathematical method developed by James Bardeen at the University of Washington because it automatically tracks combinations of physical quantities that have the same value for all observers. We spent the next months applying Bardeen's method to follow the evolution of fluctuations from the beginning of inflation up to the present, based on the Higgs field model that Andy and I had studied. The trickiest part was calculating what happens to the fluctuations as inflation comes to an end and the universe reheats to a high temperature. The Bardeen method was a surefire way for finding the right equations that follow the fluctuations through this dramatic transition, but solving those equations was technically challenging. By June, though, we reached a tentative answer and wrote a draft paper that we passed around to a few other theorists. In it, we reported that the perturbations had a scale-invariant noise pattern, as we had hoped, but the density variations were far too strong to make galaxies with a distribution that agrees with astronomical observations.

But we were not the only ones pursuing the issue. By the summer of 1982, it had become *the* red-hot topic. In June, just a few months after the papers by Linde and by Albrecht and me had appeared in print, Stephen Hawking organized the Nuffield workshop (named after its donor) at Cambridge University that, by chance, brought together many of the theorists from around the world considering the problem of density perturbations.

Turner and I were aware that Hawking was studying the problem. In fact, a few weeks before the Nuffield workshop, Hawking had given a preview of his results at the end of a talk at the Institute for Advanced Study in Princeton that I attended. He also reported a scale-invariant spectrum, in agreement with what Turner and I were finding. But Hawking also claimed that the density perturbations had the ideal strength needed to match the observed galaxy distribution. He did not explain how he obtained his answer in his talk, though, so it was not possible to understand why we were getting different answers. Like us, Hawking came to the Nuffield workshop with a draft paper that he distributed to the participants.

Alexei Starobinsky arrived from the Russian Landau Institute for Theoretical Physics with a draft paper, as well. He, too, claimed a scale-invariant spectrum, but concluded that the perturbations were somewhat stronger than what Hawking was finding.

And then Guth, working with So-Young Pi of Boston University, arrived at Nuffield ready to explore yet another approach. Guth had first become interested in the problem following Hawking's talk at Princeton. He did not attend the talk, but I had called him to describe Hawking's provocative claim, which inspired him to try the calculation himself. Since he and Pi had just started, they did not have an answer ready by the beginning of the workshop. But Guth had settled on the mathematical approach he would take, and he came to the Nuffield meeting with reams of notes, prepared to complete his calculation during the course of the meeting. (Pi did not attend the meeting.)

So, the workshop began with three draft papers with different answers, and four distinct mathematical approaches, some spelled out and some not. Despite the chaos it caused at the outset of the meeting, having the different approaches proved to be important for cracking the problem, because the calculation was so subtle and complex that mistakes were easy to make no matter which method was used. Also, each method was approximate, to some degree, invoking various simplifications and assumptions that made the calculation tractable. It was essential to have alternate methods to see if the different approximations gave consistent answers.

James Bardeen, who had pioneered the mathematical approach Turner and I were using, was also at the Cambridge meeting. He was intrigued that the theorists were getting different answers.

He was also very excited to discover that Turner and I were using his method. Turner and I agreed to collaborate with him in checking the results and completing the project.

Everyone has his own story to tell about those grueling and magnificent weeks at Nuffield, and our experiences and memories are probably not all the same. My own recollection is that the truth was reached in fits and starts, through intense interaction and cooperation among the different theorists, only really emerging in the last days of the meeting. For the first week or so, as Bardeen checked our calculation, Turner and I grew increasingly confident that we had obtained the right answer in the draft paper. We were not too concerned that Starobinsky and Hawking were each getting different results because there was nothing to suggest that they had properly tracked physical quantities that have the same value for all observers, as Bardeen's method did automatically, and this could account for the discrepancy.

We were more disturbed by the fact that Guth, carrying through a method that he shared with us, appeared to be getting a different answer. Although his results were tentative, he seemed to find that the density perturbations were much stronger than we had found, producing a universe with far too many galaxies and too much homogeneity compared to what is observed. It was difficult to compare our calculations directly. Guth used a shortcut to cover the period from the end of inflation through the reheating of the universe that worked only if one assumed that the rate of inflation is constant all along. The assumption corresponds to choosing an energy curve for the Higgs field in the shape of a step, an absolutely flat plateau ending with a sharp drop-off. By restricting himself to this special case, Guth was able to get an answer without using Bardeen's cumbersome method, but at the cost of an assumption that is physically unrealistic. Higgs field energy curves always change smoothly so that inflation must slow down gradually. None of us was sure how the slowing down of inflation affected the final answer, and Guth himself kept rechecking and comparing with us to the very end of the meeting.

And soon Bardeen, Turner, and I had our own worries. About midway through the meeting, Turner was set to present the results of our draft paper when, at around midnight the night before, Jim Bardeen knocked on his door to inform him that he had discovered a possible flaw in our calculation: one seemingly innocent mathematical approximation out of the many steps. The step needed to be replaced with a more reliable approximation before we could be sure of our answer. So, the next day, Turner had to present a rather tentative result with a promise to get a final answer by the end of the meeting. From that point until the end of the meeting, the three of us got very little sleep.

In the meantime, when Starobinsky gave his presentation, he reported an answer that had shifted a bit and was now in rough numerical agreement with what Guth was telling us. However, as in his paper, he did not explain enough about his method of calculation for us to check against ours and to identify the source of the disagreement.

Then, Hawking gave a talk in which he changed his earlier answer. Now he, too, claimed the fluctuations were too strong, in agreement with Guth's tentative result and Starobinsky's claim. Hawking did not spell out his method; nor did he explain why the answer had changed.

This placed the burden squarely on our shoulders. We were the only group using a rigorous method, and ours was the only mathematical approach that could be adapted to realistic cases in which there is a smoothly changing energy curve and the rate of inflation slows gradually at the end. Finally, just before the end of the workshop, Bardeen, Turner, and I identified a reliable mathematical short-cut to replace the questionable step that Bardeen had identified. And, sure enough, our result substantially changed to one that was now in qualitative agreement with the others in the idealized case they considered, in which the inflation rate is constant. But, now that our method was completely worked out, we could also apply it to realistic cases, where the other methods could not be applied at all, and show that the outcome was similar. (The method we used is now the standard approach cosmologists use for making inflationary predictions.) Ironically, the draft paper that Turner and I had prepared before Nuffield had been very nearly right all along, including all the correct equations to be solved. If we had tried to solve the equations by computer rather than by hand using shortcuts, we would have obtained the right results months earlier.

By the end of the three weeks, when Guth presented his final result, all four groups reported that their answers had converged. Finally, all the groups could be secure that the problem had been licked. We all left Cambridge exhausted but exhilarated by the remarkable agreement we had achieved and aware that history had been made.

The exciting conclusion emerging from the Nuffield workshop was that there is, in principle, a proven mechanism for creating energy density variations after inflation through the quantum fluctuations of the Higgs field as it is rolling down an energy curve. The variations have a scale-invariant noise pattern, like the one illustrated on page 57, that agrees with astronomical observations. The mechanism is surprisingly simple: quantum fluctuations occur randomly in space and time and kick the Higgs field a tiny way up or down the energy curve. Kicks downhill bring the field closer to the precipice and end inflation earlier than average. Kicks uphill delay the end of inflation. Whenever inflation ends, the energy stored in the Higgs field converts to hot matter and radiation, which begin to expand and cool. Because inflation ends at different times in different places, the temperature and density vary from place to place as well. Unknown to us at the time, some of these ideas had also been suggested independently by William Press at Harvard University and by V. (Slava) Mukhanov and G. Chibisov at the Lebedev Physical Institute in Moscow, although without an accompanying reliable method for calculating the strength of the perturbations in realistic models.

Bardeen, Turner, and I particularly emphasized an additional noteworthy prediction: for realistic models, we found that the pattern is *not quite* scale-invariant. If one views the density pattern as a sum of sinusoidal waves, as shown on page 57, then, instead of having the same height, the waves become gradually smaller as the wavelength decreases. To call attention to the point, we entitled our paper "Spontaneous Creation of *Almost* Scale-Free Density Perturbations in an Inflationary Universe." At the time, most astronomers considered the deviation from exact scale-invariance to be an academic issue, too tiny to consider seriously. But, today, experiments like WMAP can actually measure it, and the prediction is considered one of the milestone tests of inflation, as will be discussed in chapter 9.

As the news of the Nuffield meeting began to spread, physicists seemed to be enormously impressed. Inflation had not been designed to produce a scale-invariant pattern of energy density fluctuations, and yet it did. It is not unusual for a theory to produce an unintended result that is wrong; that is a common way that theories are disproved. It is very rare for a theory to produce an important unintended result that is correct. When this occurs, physicists consider it a strong sign that the theory is likely to be right.

The victory gave theorists more confidence in the inflationary prediction of flatness. Since it was known that ordinary matter accounts for only 5 percent of the energy density needed for a flat universe, according to Friedmann's equation, this suggested that the other 95 percent must be dark matter. (Few considered dark energy a serious possibility at the time.) This suggestion seemed to fit perfectly with the mounting evidence from astronomers indicating the existence of halos of dark matter around all galaxies. Many interpreted this as another triumph for inflation.

Of course, there was also bad news emerging from the Nuffield workshop. The four groups agreed that the density variations produced by Higgs fields are scale-invariant, but also found that they are much too strong. The prediction for the cosmic background radiation (that is, the WMAP image), for example, was that the temperature difference between the hottest and the coldest spots across the sky should be several degrees. In 1982, this prediction was already known to be wrong. There were measurements of the cosmic microwave background showing that the temperature difference is much less than one degree, although exactly how much less had not been determined. A decade later, the COBE satellite measured the actual temperature difference to be over ten thousand times smaller than the theoretical prediction for Higgs-driven inflation.

While the failure was a setback, it seemed surmountable. Turner, Bardeen, and I, for example, immediately showed that it was possible for other kinds of quantum fields to have energy curves of the right shape. We also showed how the candidates might be distinguished by measuring the small deviations from exact scale-invariance. By the spring of 1983, Turner and I had extracted from those examples a clearly defined set of conditions that a quantum field must have to be a viable candidate for the "inflaton," the generic name for the field that drives inflation. Our rules were simple enough that they could fit onto tiny slips of paper the size of medical prescriptions. To emphasize the point, we made joke medical "prescriptions" (including an ℞ masthead), which we brought to the first

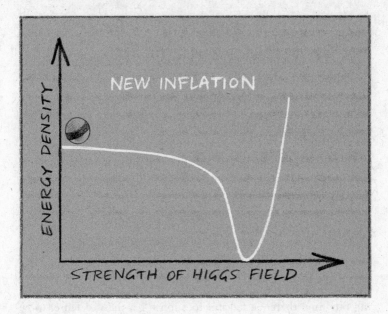

major post-Nuffield gathering of cosmologists that was set to take place in Aspen, Colorado, that summer.

As I drove cross-country with my family from Philadelphia toward Aspen, I wondered what the meeting would be like. I was still a novice in cosmology, having worked in the field for fewer than three years. My previous experiences at cosmology meetings, with the exception of the Nuffield gathering, were mostly conferences with little chance for exchange or criticism. The Aspen workshops were said to be organized with plenty of time for questions and interaction, but this was my first time at Aspen, and I was not sure how exactly this would work in practice. A broad spectrum of astrophysicists and cosmologists, as well as particle physicists, would be there. Having had over a year to digest the new inflationary models that Linde, Albrecht, and I had proposed and the results from the Nuffield workshop, would the community come together and embrace or reject the new ideas? I really did not know.

With so many senior physicists whom I had never met before scheduled to be there, I was also looking forward to making new contacts and perhaps finding new collaborators. What I did not anticipate was that my most fateful encounter would be with the youngest participant at the workshop, a fellow from Imperial College, London, who did not even have a formal Ph.D. degree (which, technically, violated the official rules for participating in an Aspen Institute workshop). I already knew him from a distance because I had read his Ph.D. thesis.

I had been asked to be the outside examiner on Neil Turok's Ph.D. oral defense committee, the group of senior physicists who grill the candidate on his research and then judge whether the work is worthy of the degree. Because both Neil and his adviser, David Olive, were attending the Aspen workshop, Neil's oral presentation was arranged to take place at the Aspen Institute on some afternoon during our time there—a fanciful location for a thesis defense, to be sure. Although Neil's thesis was mostly mathematical in nature, with only one section somewhat related to cosmology, I decided to focus on that portion during the oral presentation. I wanted to test whether he had any serious interest or talent in this area. I was very impressed by the outcome. Although he was obviously new to cosmology, Neil displayed an unusual combination of technical prowess, creativity, and self-confidence. I passed him, of course. But in addition to that, I made a mental note to follow this talented fellow's career and look for an opportunity to collaborate with him in the future. Since he immediately headed off to develop his own ideas, based on his thesis, which competed with inflationary cosmology and with my own work, the opportunity for collaboration did not present itself immediately. In fact, fifteen years passed before the chance arose. But when it finally did, I jumped at it, realizing that here was a perfect partner with the skill and willingness to take chances on a radical new approach to cosmology.

Neil's Story.

My introduction to fundamental physics also came during a lecture. I was in the final year of my undergraduate degree at Cambridge when, on April 29, 1980, Stephen Hawking gave his inaugural lecture as newly appointed Lucasian Professor of Mathematics, the post once held by Sir Isaac Newton. The title was provocative: "Is the End in Sight for Theoretical Physics?" Even more provocative was the answer: yes. Hawking claimed that the theory of supergravity (discussed in chapter 6) would ultimately provide the unified "theory of everything" that had been the goal of fundamental physics since Einstein. While I was skeptical of so bold a claim, I was sufficiently intrigued that I began to seriously consider taking a Ph.D. in theoretical physics. I never imagined that I would eventually become a close colleague of Hawking's, and would even join him in several research projects.

I went to Imperial College in London for my Ph.D. because I wanted a change from Cambridge and I had heard good things about the course and the faculty there. Theoretical physics is a highly technical field, and as a fresh graduate, one relies very strongly on the advice of senior colleagues as to which lines of research are the most interesting and promising to pursue. My adviser, David Olive, was already well known as a brilliant and original mathematical physicist. He set me to work on a highly mathematical project concerning the behavior of fields like those responsible for the strong, electromagnetic, and weak forces. These fields are governed by very simple and beautiful equations, but the equations are very hard to solve, and they are especially difficult to handle when quantum effects are included. Olive was exploring new approaches based on deep symmetries, and he drew me into this work.

Near the end of the project, concerned that my research was

too formal and abstract to connect with the real world, I wandered into the office of one of my professors, Tom Kibble, to express my frustration. Kibble is one of the United Kingdom's most distinguished theoretical physicists. He had been thinking about an entirely new way of testing unified theories of particle physics using cosmology. At the time, I knew nothing about cosmology: there were no courses on it at either the undergraduate or the graduate level. So it came as a revelation to me that some of the very pure ideas involved in building unified field theories of particle physics might have rather direct consequences for the universe. Most exciting of all was the possibility of testing for these effects through astronomical observations.

A few years earlier, Kibble had realized that many unified theories automatically predicted that objects called *cosmic strings* would form in the extreme conditions of the hot early universe. Cosmic strings are thin strands of concentrated energy that crisscross space in a spaghetti-like network and progressively straighten themselves out as the universe expands. Kibble and others had speculated that it might be possible to see these cosmic strings through careful observations, and that they might even have played a role in the formation of galaxies. Intrigued by this potentially spectacular link between fundamental physics and cosmology, I started working with Kibble on cosmic strings while finishing my main Ph.D. project.

Soon after, at a summer school in Erice, Sicily, in 1982, I met Andy Albrecht, the same graduate student who'd worked with Paul on developing the first practicable model of cosmic inflation. Albrecht and I became close friends and remained in contact over the following year as we each completed our Ph.D.s. We decided to apply to the Aspen Center, whose program for 1983 featured a workshop on the interface between particle physics and cosmology, the same workshop Mike Turner and Paul were attending. To our delight, Andy and I were both accepted. We met in Washington, D.C., to take a Greyhound bus together to Aspen, so that we could see a bit of the country while having lots of time to discuss physics along the way.

For me, these discussions were a revelation. In the United Kingdom, theoretical physicists tended to specialize in narrow topics and pursue rather well-defined paths. But in the United States, I discovered, the spirit was much more freewheeling. Andy represented the American tradition, most famously exemplified by Richard Feynman, in which the whole point of doing theoretical physics is to figure out everything for yourself, in your own way, from scratch. As we traveled on the bus day and night toward Colorado, discussing quantum physics, statistical physics, inflation, and more, we wondered what the renowned Aspen Center for Physics would hold for us. Was it really possible to apply quantum physics and unified theories to the universe? Was the whole subject a fantasy? Or were we witnessing the start of something really big?

At a small town in the Midwest, a gentleman about seventy years old got on the bus. Soon he started excitedly telling everyone he was going to Las Vegas. He had withdrawn his life savings and was going to gamble it all in a last-ditch attempt to make it rich. The two of us laughed at how naive he was, but in retrospect, we weren't so very different. We were pinning our hopes on new and incredibly ambitious lines of research.

The Aspen Center for Physics is specifically designed to promote innovative research and collaborations. A chic little town nestled high up in the Rocky Mountains, Aspen also has a strong hippie streak. Once during the workshop, I answered the phone at the center only to find the caller earnestly asking, "Is this the Aspen Center for Psychics?" The question seemed oddly appropriate. The workshop I was attending was attempting to divine some of nature's deepest mysteries, albeit using mathematics and physics rather than a Ouija board.

At the workshop, everyone was excited about the new connections between particle physics and cosmology. The theory of quantum generation of density perturbations during inflation had been developed the previous summer, and many consequences were still being worked out. Paul and Mike Turner gave a seminar at which they handed out their joke medical prescriptions for constructing successful inflationary models, each signed by "Steinhardt and Turner, Doctors of Inflatology." It was evident that inflation had gained an enormous number of followers and was being rapidly accepted as part of the standard model of cosmology.

In the spring, David Olive had written to Paul asking him to serve as external examiner for my thesis, and Paul had agreed. Since all three of us were planning to be in Aspen that summer, it made sense to hold the thesis defense there. Most of the thesis was devoted to mathematical physics, with only a short section on cosmic strings. But the defense focused entirely on that section, especially the prospects for testing the notion that galaxies might have formed around string loops. This proposal was much less ambitious than inflation, since it did not attempt to explain the smoothness and flatness of the universe nor how it had emerged in a hot, dense state. Instead, these things were just assumed and we asked how strings might then form as the universe cooled, and later stir up the matter, thereby generating structures like galaxies. The idea was attractive because it was based directly on the fundamental notions of unification and symmetry breaking. It was more predictive than inflation because the behavior of the strings is very insensitive to the details of the Higgs energy curve. Finally, there was the exciting possibility of detecting a cosmic string, which would be an unmistakable remnant of grand unification in the early universe. This possibility is still of great interest, and from time to time there are reports of the detection of cosmic strings, although so far none has been confirmed.

The Center for Physics provided a relaxed atmosphere where people could work, talk, and just hang out. There were a few formal presentations, but most of the work was actually done during informal discussions. Away from their normal responsibilities, people developed new ideas and established new collaborations. At lunchtime, everyone joined in volleyball games, but even here, people on the sidelines waiting for their turn to play discussed physics. On the weekends, groups returning from long hikes or bike rides in the mountains often came back with new ideas. It sounds a bit like a holiday camp, but the scientific focus is intense, and the center has been the site of countless key innovations over the years.

All sorts of exciting new ideas were being discussed in the summer of 1983. Many theorists were trying to develop better models for inflationary energy. Others were starting to perform giant computer simulations of how the density fluctuations created at the end of inflation draw together dark matter particles of various types to form galaxies. The idea of extra dimensions was beginning to attract attention, and some researchers were beginning to wonder what difference they might make to early universe cosmology. Everyone listened and participated as many new ideas were born, and almost as many died quick deaths.

Still, the mood was one of huge optimism. The new discipline of particle cosmology was emerging before our eyes, a discipline that would set the agenda for the next two decades of discovery about the universe. We could all see that the excitement was drawing a combination of new ideas and talent into astrophysics and cosmology. As the famous U.S. astronomer Vera Rubin would put it thirteen years later at a meeting in Princeton, particle physicists were traditionally the "linebackers" of physics. By 1983, bringing these heavy hitters into the field was already shifting the priorities in astronomy toward using the universe as a giant testing ground for fundamental physics.

More important, it was revolutionizing views about the history of the universe.

That summer at Aspen, it was already clear that inflation was set to become the dominant cosmological theory. While I was tremendously impressed with the achievements of the inflationary theorists, I still had my doubts, perhaps because of my background in pure theory. The inflationary energy curve seemed contrived and ad hoc. And why did the field whose stored energy drove inflation, the "inflaton," begin up the hill, on the high energy plateau, rather than at the point of lowest energy where it ultimately settles? In the original picture of Higgs inflation, the starting point was explained. As the universe emerged from the big bang, it would be filled with a hot plasma whose high temperature would force the Higgs field to zero. But Higgs field inflation did not work. According to Steinhardt and Turner's detailed prescription, a successful inflationary model required the inflaton to interact so weakly with the plasma that it would be completely unaffected by the high temperature. So this explanation for why the inflaton started out up the hill fell away. Some theorists, like Andrei Linde, argued for "chaotic" initial conditions, according to which the inflaton field would just be randomly distributed across space in the infant universe. The idea was that regions where the inflaton happened to lie on the plateau would undergo inflation and would produce a universe looking like ours. I found this picture vague and unconvincing because there was no theory to explain the initial "chaos": it was just put in by hand. Furthermore, the universe could not be completely chaotic on all length scales or it would be too nonuniform for inflation to start. Something was needed to explain this initial mixture of chaos and order, but it seemed to me that the solution being proposed was, in effect, just to stop worrying about the problem.

Cracks in the foundations of grand unification theory were also beginning to appear. The strong, weak, and electromagnetic forces did not quite merge at high energies, as the simplest grand unified models predicted. And some of the predictions for the masses of matter particles, like electrons and quarks, came out wrong. The most dramatic prediction of grand unification was that protons and neutrons, the basic constituents of atomic nuclei, should be able to decay into lighter particles. The average decay time is very long: 10^{30} years, or longer than the current age of the universe. But in a large amount of material—a ton of water, for example—there are a huge number of protons and neutrons that will decay sooner. According to the simplest grand unified theories, if you observed this quantity of water for several years you would see a few protons decay, in a detectable burst of radiation and particles. Beginning in the 1980s, a number of experiments were constructed to search for this process, but to this day, not one decay has been seen. Theorists could modify the simplest grand unified models to evade the problem by introducing more ad hoc Higgs fields to slow down the decay. But this made the models more complex and less attractive.

After my thesis defense in Aspen, there was champagne all around. As all of us celebrated, we speculated about where the new field of particle cosmology might lead. The growing problems in the theories of inflation and grand unification were worrisome, but the mood was nevertheless sanguine. Many anticipated that the setbacks would be minor and that particle physicists, cosmologists, and astronomers would henceforth work together in a powerful, combined discipline that would advance our knowledge of, simultaneously, the very small and the very large. And although Paul was headed back to his faculty position at the University of Pennsylvania and I was beginning a new postdoctoral fellowship at the Institute for Theoretical Physics in Santa Barbara, California, we both felt that we were likely to work together on some common project in the near future.

The Aspen dreams turned out to be wrong on nearly all counts. Over the next fifteen years, the two of us pursued different scientific directions. Paul concentrated on developing inflationary theory, while I focused instead on testable consequences of symmetry breaking in the early universe.

Over the same period, cosmologists and particle theorists went their separate ways, as well. The cosmologists began applying the new elements that had emerged from the considerations of the early universe, such as inflation, cosmic strings, and dark matter, to explain the formation of galaxies and to predict the temperature variations in the cosmic background radiation. With advances in technology, they realized, the exciting new ideas could actually be tested. As for the particle theorists, they were about to abandon field theory and grand unified theories to pursue a revolutionary new direction known as string theory, which would so envelop them with mathematical challenges that cosmology would be set aside. Meanwhile, theoretical physicists began to worry whether particles, fields, and grand unification formed the right approach after all. And if the basic approach to the fundamental laws of physics had to be changed, I wondered, could another revolution in cosmology be on the horizon?

In 1988, I was offered an assistant professorship at Princeton, a world center for both string theory and cosmology, and I jumped at the chance of working to connect the two. By then, string theory had superceded supergravity as the leading contender for a unified theory, although the two would later be merged into a grander framework called M theory. According to string theory, every known particle is actually a tiny piece of vibrating string, with the string vibrating in a different way for each different type of particle. As well as tiny pieces of string, it is possible to have very long lengths of string, which behave just like the cosmic strings I had been studying. One of the first things I did at Princeton was to work out exactly how these long strings could have emerged from the hot early universe. But the string theorists at Princeton were not very interested in this type of practical question. They preferred to focus on more formal developments, which they hoped would uncover deeper theoretical principles underlying string theory. Try as I might to interest them, they told me it was too early to consider these cosmological problems.

At the same time, I had great fun talking with the cosmologists at Princeton and generally making the most of the university's stimulating atmosphere, where such illustrious physicists as Albert Einstein have worked. While I followed the formal developments in string theory with interest, I busied myself with a range of simpler cosmological puzzles, such as the general consequences of symmetry breaking in the early universe, the reason for the preponderance of matter over antimatter in the universe, and the polarization of the cosmic microwave sky, as well as performing observational tests for a cosmological constant. My work progressed well, and I was eventually made a full professor at Princeton before I decided to move back to the United Kingdom, in 1996, for family reasons. Soon after I left Princeton, Paul took up the Albert Einstein Professorship there, so, unfortunately, we never overlapped.

As I arrived in Cambridge, the race to measure the fine detail in the temperature variations of the cosmic microwave sky was starting to heat up. With the help of bright young collaborators, powerful new computer simulations, and clever mathematical techniques, I was able to complete the work Albrecht and I had initiated nearly a decade before, aimed at calculating the pattern of temperature variations that should be seen if cosmic strings, or other similar structures formed by symmetry breaking in the early universe, had really been responsible for galaxy formation. When the measurements were made, they conflicted with our predictions, and the idea was disproved. It was of course sad to see such a simple and beautiful idea fail, but it was also real progress and helped to build the growing consensus around the inflationary model. I still felt nervous about inflation; I thought the model needed firmer foundations. In particular, we still had to explain why the inflaton field started out high up on the energy curve. At Cambridge, I had the opportunity to learn at first hand about an elegant proposal made by Stephen Hawking, with James Hartle of the University of California at Santa Barbara, for the initial state of the universe. Working with Stephen, I calculated what the Hartle-Hawking proposal meant for the inflaton field. Unfortunately, the result was nothing like Linde's expectation based on his intuitive chaotic inflation picture. According to Hartle and Hawking's formula, the inflaton field started out near the bottom of the energy curve, in the *lowest* energy state, meaning that there would be no inflation after all.

So it was with a very open mind that I decided to propose a scientific program at the Isaac Newton Institute for Mathematical Sciences in Cambridge devoted to cutting-edge issues in cosmology. At this time, observations were making huge strides and, obviously, a major focus of the workshop would be making sense of all the new data. But the program would also be an important forum for discussing new and improved cosmological theories. They continued to provide a vital stimulus to the field, and, of course, they are ultimately essential if we intend to really understand the universe, as opposed to merely describing it. To develop the proposal I needed some co-organizers, and it didn't take me long to decide who the ideal choices would be. I had known Paul for fifteen years, during which he had become one of the preeminent exponents of realistic inflationary models and, in particular, the successful version of the inflationary model that included dark energy. What distinguished Paul from other experts, however, was his unusually broad grasp of physics, his openness to new ideas, and his energy in exploring them. The combination has allowed him to make key innovative contributions in many different fields of physics over the years. A second ideal organizer also immediately sprang to mind. I had come to know Valery Rubakov, of the Institute for Nuclear Research in Moscow, through my theoretical work attempting to explain why there is more matter than antimatter in the universe. Valery represents a very strong Russian tradition in theoretical physics, dating back to the great Lev Landau. He is expert in formal aspects of quantum fields and gravity, but he is also interested in cosmology and open to new ideas. An added bonus of having Paul from the United States and Valery from Russia would be the ability to identify and attract the world's leading researchers from both East and West.

To my delight, both Paul and Valery accepted immediately. We decided to kick off the program with two conferences: one on the latest developments on the observational side, and one on new theoretical ideas. Both turned out to be ideally timed. A slew of new observations preceded the first conference, confirming the current inflationary picture and ruling out all of the main rival models. As a result, the meeting was the first international gathering where there was truly a consensus on the broad features of the universe.

The second meeting was also very fortunately timed. In the late 1990s, string theory had undergone several major developments. It had been combined with supergravity into a new theory called M theory, which included new objects called *branes* (short for "membranes") and led to a new picture of extra dimensions of space. In the year leading up to our meeting, there had been a burst of activity in various directions, attempting to use these new features to build realistic models of the universe. In the final month as the meeting approached, more and more people signed up to attend. The meeting was a huge success, with excitement (and the audience) building over the course of the week and many participants telling us it was the best they had ever attended.

The Isaac Newton Institute meeting was, above all, a vital stimulus to our own research. It convinced us that, finally, string theory and supergravity had something really interesting and new to say about the cosmos. After more than a decade, Paul and I had finally converged on a project we wanted to pursue together.

chapter six

From Strings to Ekpyrosis

Make everything as simple as possible but not simpler.

—Albert Einstein

Nowhere was the optimism of particle physicists in the early 1980s more evident than at the annual Workshop on Grand Unification, known by the acronym WOGU (pronounced "whoa-goo"). Each spring the leading physicists, their postdoctoral fellows, and their students would gather at a different site to discuss the latest experimental breakthroughs and theoretical advances. Every year, the exciting presentations at WOGU seemed to engender new confidence that quantum field theory and grand unification were on track . . . until the fourth WOGU, when a soft-spoken young theorist politely suggested that a sharp turn in the current thinking might be needed.

The meeting took place in April 1983 at the University of Pennsylvania, in Philadelphia, about fifty miles from Princeton, New Jersey, the home of Edward Witten. Only thirty-two years old at the time, he was already recognized as a theoretical physicist of great vision. For years, he had been a much admired pioneer in exploring the theoretical underpinnings of grand unification.

Paul was a professor at the University of Pennsylvania at the time and, as one of WOGU's organizers, had been assigned to call Witten to invite him to give a presentation. Surprisingly, Witten was reluctant to accept. He explained that he was working on something new and was not sure the topic would be appropriate for a meeting on grand unified theories. That only made the prospect more intriguing, so Paul persisted and Witten finally agreed to speak.

When the time came for Witten's talk, the last of the meeting, the auditorium was packed to standing room only. In his characteristic calm and gentle voice, Witten began by noting ways in which the current attempts at grand unification were failing. The most dramatic prediction, the instability of protons, had been tested, but no decays had been seen. The predictions of the masses of matter particles had also turned out wrong. Physicists could adjust the models to evade these problems, but only at the cost of adding ugly complications that made the whole framework implausible.

Witten then suggested that it might be time to consider a totally new approach. He proposed three guiding principles. First, the new approach should include gravity from the outset. Particle physicists were used to ignoring gravity because the gravitational attraction between elementary particles is normally negligible. However, when particles are smashed together at high energies, their collective mass rises in accordance with Einstein's famous equation $E = mc^2$, and the effects of gravity become stronger and stronger. At the very high energies where the strong and electroweak forces seem to merge into a single unified force, gravity is nearly as strong. For this reason, Witten argued, gravity has to be included in any theory of unification.

Dealing with gravity would be no easy task. Einstein had developed his theory of gravity in the early part of the twentieth century, at the same time that quantum theory was emerging. Despite all attempts, the two strands of physics had never been successfully joined. Einstein's theory works tremendously well on large scales for describing gravity on the Earth, in the solar system, and in the universe. But just like electromagnetism and light, gravity must be formulated in a way that is consistent with the laws of quantum physics in order to make sense on microscopic scales. For the other three forces, the quantum field approach had been spectacularly successful. But for gravity, every attempt to quantize Einstein's theory had failed, leading to infinities, negative probabilities, or, at best, an infinite number of indeterminate parameters. A totally new approach was needed, one that would give a sensible answer.

Everyone in the audience knew about these difficulties in building a quantum theory of gravity. So we were all naturally anxious to learn what Witten had in mind. Witten emphasized that he did not deserve credit for the idea he was going to suggest. Hard work had been done by a small, intrepid group of theorists working largely unnoticed and unappreciated. But Witten was now advocating, as his second principle, considering their daring proposal: a conceptual framework known as string theory.

Many in the auditorium had heard of string theory before, but most knew little about its history because it had had little impact on mainstream particle physics or cosmology up to that point. String theory had been developed in a rather roundabout way.

In 1968, Gabriele Veneziano at the European Organization for Nuclear Research (CERN) had proposed a formula for describing the scattering of nuclear particles interacting via the strong nuclear force. In 1970, Yoichiro Nambu at the University of Chicago, Holger Nielsen at the Niels Bohr Institute in Copenhagen, and Leonard Susskind, then at Belfer Graduate College in Israel and now at Stanford University, showed that Veneziano's formula could be interpreted as a model of vibrating one-dimensional strings. Unfortunately, it was soon discovered that the model had various pathologies, such as a *tachyon*, a physically impossible particle that moves faster than light. But this problem was cured as people realized that string theory was much more than a theory of nuclear particles. First, Joël Scherk at the École Normale Superieure in Paris and John Schwarz at the California Institute of Technology showed that string theory included a particle behaving like a graviton, the troublesome quantum of Einstein's theory of gravity. Then, by incorporating matter particles using a powerful new quantum symmetry called supersymmetry, Scherk with David Olive (Neil's Ph.D. thesis adviser) and other physicists managed to construct a completely consistent model with no tachyon.

In this way, the theory originally designed to describe the strong nuclear force was suddenly transformed into a unified theory with the potential to describe all the forces and particles in nature, including quantized gravity. But these developments went largely unnoticed. The 1970s were the heyday of quantum field theory, and string theory was seen as a speculative backwater. A few lonely theorists continued to struggle to develop the theory and iron out its remaining mathematical difficulties. This was a daunting and slow process, since few people were willing to risk working on the subject.

Witten's talk went on to describe the advantages of reinterpreting elementary particles as tiny spinning bits of string. Just as Einstein pictured three-dimensional space as an elastic substance that can be stretched and distorted, you can think of string as a geometrical curve with no width that can bend and turn in all possible ways, like an infinitely thin strand of rubber. The string is perfectly elastic, so it can shrink away to a point or be stretched out to an arbitrary length.

If you stretch a piece of string out in a straight line, the free ends pull together with a fixed force called the string tension.

Some of the properties of string are actually very similar to those of cosmic strings, which were mentioned in the last chapter. But whereas cosmic strings are really twisted-up configurations of fields with a minuscule but finite width, fundamental strings are ideal one-dimensional mathematical curves.

The string picture is beautiful in that one basic entity—string—can potentially account for the myriad of elementary particles observed in nature. Bits of string vibrate and spin, in certain specific quantized motions. Each new quantized state has a set of physical attributes: mass, charge, and spin. The little pieces of string describing photons, electrons, or gravitons are far too tiny to be seen, much less than a trillionth the diameter of a proton. To us, they appear like pointlike particles. But if string theory is correct, the masses, charges, and spins of these little bits of string should precisely match the physical properties of all of the particles ever discovered.

Witten was especially attracted to this picture because it included gravitons as a hidden bonus, as Scherk and Schwarz had first shown. Bits of string with two free ends could account for all known types of matter particles. But the mathematics of string also allows for closed loops, like tiny elastic bands. When vibrating and spinning in just the right way, these loops have the same properties as gravitons, the quanta of the gravitational field. Even better, while calculations assuming pointlike particles and gravitons give nonsensical, infinite answers, calculations for stringy particles and loopy gravitons produce sensible, finite results. Although not designed for the purpose, string theory appears to automatically incorporate a theory of quantum gravity without infinities.

The reason string theory works where the particle description of quantum field theory fails can be explained by simple geometry. If two pointlike particles collide, their energy is concentrated at a point. Such pileups of energy cause a large gravitational field, curving space and drawing even more energy into the region. A runaway process ensues in which space curls up irretrievably into a tinier and tinier knot: a singularity. This catastrophe leads to mathematical infinities signaling a breakdown of the theory. On the other hand, if particles are tiny vibrating strings, their energy is spread out. If a collision causes a momentary pileup of energy, the string rapidly wriggles away and spreads out the energy, preventing the gravitational distortion from concentrating in one spot. Calculations of what happens when two bits of string collide, join, and break apart again give sensible, finite results. There are no singularities, and no infinities.

Witten's third guiding principle dealt with the major hitch theorists had previously discovered about string theory. The equations describing the quantized vibrations of strings give sensible answers only if the number of spatial dimensions is nine. Nine?! To most physicists, this seemed absurd. Why study a theory that predicts six extra dimensions of space that have never been seen?

Witten addressed the problem of extra dimensions head-on: Learn to live with them, he said. Just accept the six extra dimensions of string theory; they are an essential aspect of the geometry of the universe. He reminded the audience that back in the 1920s the Swedish physicist Oskar Klein, building on the work of the German physicist Theodor Kaluza, had dreamed up a way of linking Maxwell's electromagnetic theory with Einstein's theory of gravity, in a model of the universe where one extra dimension of space was hidden from view.

To see how this works, consider the surface of a long soda straw. From a long distance away, it appears to be one-dimensional because you cannot detect its thickness. But up close, you can see the surface of the straw. To prove to yourself that the surface is two dimensional, slit the straw along its length and flatten it out. You will get a rectangle, a shape that is obviously two dimensional because it has both length and width.

Klein supposed that in addition to the three familiar dimensions of height, width, and length, there is a fourth dimension of space that is curled up in a circle so tiny that it cannot normally be seen. Kaluza and Klein's remarkable discovery was that Einstein's theory of gravity in four space dimensions, with one of the dimensions curled up as described, contained both Einstein's theory of gravity in the remaining three extended dimensions *and* Maxwell's theory of electromagnetism. Electric and magnetic fields arise, in this picture, from a "twisting" of the small extra dimension as you move along one of the large everyday dimensions.

According to Witten, theorists simply had to adapt Klein's idea to the six extra spatial dimensions in string theory. There is no prob-

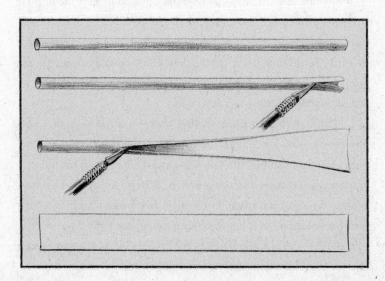

lem having strings wiggle in nine spatial dimensions, so long as six of the spatial dimensions are too small to be seen.

The extra dimensions would exist at every point in three-dimensional space. As an analogy, consider a pile carpet made of woolen loops. To us, looking from above, it appears as a two-dimensional surface. But to an ant, it seems like a huge forest of loops. At any point, the ant can choose to run along the direction of the floor, that is, along one of the two extended dimensions, or around one of the woolen loops that describe the curled-up dimension. In the same way, the extra dimensions in Kaluza and Klein's approach are invisible, because their tiny size is too small to be seen. But in principle, with a very powerful microscope using very short wavelength radiation, one would be able, like the ants on the pile carpet, to see the convoluted structure of the extra dimensions on tiny length scales.

Witten framed his lecture carefully and peppered it with qualifications, but his message was clear. In a mere forty minutes, he made a compelling case that theories of grand unification were incomplete and that gravity, strings, and extra dimensions ought to be considered. Research on the fundamental laws of physics could be headed toward a revolution, he quietly suggested. You could have heard a pin drop in the auditorium. The audience was stunned, unsure how seriously to take Witten's remarks.

Through the remainder of 1983, there were few signs that anything was going to change. During the Aspen summer workshop that year, for example, the talk was almost all about grand unification and field theory. But, sure enough, Witten's lecture was the harbinger of a revolution that would soon sweep the world. The "first string revolution," as it has since been called, was ignited a year later at the 1984 Aspen workshop when Michael Green, then at Queen Mary College, London (now at Cambridge), and John Schwarz overcame a key mathematical roadblock in the construction of realistic string theories.

Until that point, there were many versions of string theory with different ways of folding the extra dimensions, but they all seemed to be fatally flawed. Witten had recently shown that many versions of string theory are unacceptable because they violate the conservation of energy through a quantum effect known as an *anomaly*. Green and Schwarz's breakthrough was the identification of a special version of string theory that had realistic matter particles and no anomalies. Now, for the first time, one could point to a quantum theory that incorporated gravity and other forces and gave finite, sensible answers.

Working at Princeton, David Gross, one of the leading pioneers of unified quantum field theories (now director of the Kavli Institute for Theoretical Physics in Santa Barbara), along with Jeffrey Harvey and Emil Martinec (both now at the University of Chicago) and Ryan Rohm (now at Boston University) produced a compelling example known as *heterotic string theory*. The word *heterotic*, meaning hybrid, was added because it combined different versions of string theory to obtain one that has more of the ingredients needed to make a realistic theory of elementary particle physics. (A later, further improved form, heterotic M theory, was the stimulus for our work on the cyclic model of the universe.) These successes, and others that followed in rapid succession, captivated the international community of theoretical physicists. Almost overnight, it seemed, the focus of research shifted from particles to strings. And the merger of fundamental physics and cosmology that had seemed imminent in 1983 was put on hold.

Six-Dimensional Origami

String theory was attractive because it seemingly incorporated all of the features of particle physics and gravity within a very tight mathematical framework. Whereas there were infinitely many versions of quantum field theory with different numbers of fields and forces but no deep principle for choosing between them, there are just five known mathematically consistent string theories. Each describes a nine-dimensional world in which everything depends on only one basic quantity, the string tension, and in which spinning bits of string describe every possible type of particle. Nevertheless, because six of the dimensions of space may be curled up in very many different ways, and because the bits of string will respond accordingly, each string theory can in principle describe a huge range of models of the three-dimensional world. All of the complexity in nature, which rests on the distinctions in character between the different matter particles and force fields, might arise from the complicated geometry of the six extra spatial dimensions.

How can the geometry of hidden dimensions produce complexity? As a simplified analogy, imagine that you are given a set of electrical wires that appear identical from the outside and are challenged to explain their different properties without being able to see inside their insulating coats. Your measurements show that the electrical resistance per meter, the mass per meter, the heat resistance per meter, and so on, vary widely from wire to wire. You might guess, at first, that every wire is made of a different type of metal. If that is the case, their properties are all unrelated and you gain no deeper insight. Then suddenly it hits you: there is another possibility. Maybe all the wires are made of the same material—copper, for example— but arranged in a different geometry. For example, there might be one or more strands, or the thickness could vary from wire to wire. The differences in physical properties may be entirely due to this geometry, rather than to the composition. If this bold proposal is right, the electrical resistance, mass, and heat conductivity of all the wires should be related to one another in a predictable way. If you test these predictions and find that they always match the observed properties, you will have understood a great deal about the internal structure of the wires, without ever looking inside the insulating coats.

In a similar way, string theory can be proven correct by identifying a geometry of the unseen dimensions that makes it possible to explain the known masses, charges, spins, and other physical properties of elementary particles. Of course, this means that string theorists have to analyze all the possible shapes for the six extra dimensions and try to find which one fits the real world. This is a hugely challenging mathematical problem—it is like exploring six-dimensional origami. Nevertheless, according to string theory, this study is crucial to understanding the real world.

The Blossoming of a Geometrical Garden

As string theorists explored the full complexity of the six extra dimensions, they discovered a number of technical problems that convinced them that none of the string theories was actually complete. In addition to one-dimensional strings, there had to be membranes with two or more dimensions. These additional objects are just higher-dimensional versions of string. If you pull on a string, the force that pulls back is the string tension. Stretch out a length of string, and the energy stored in it is just the length times the tension. Now think of a two-dimensional membrane, like a piece of balloon. If it has a fixed surface tension, its energy grows in proportion to its

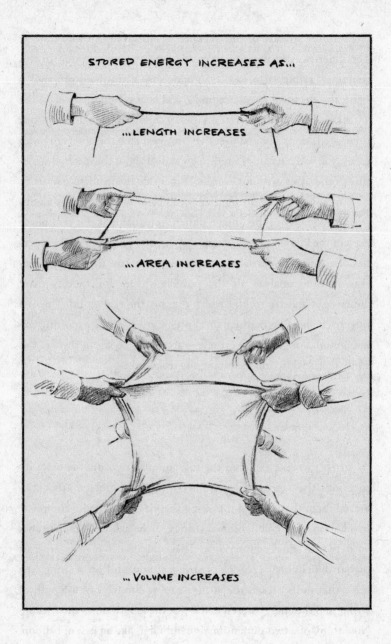

area as it is stretched out. Likewise, if you stretch out a three-dimensional membrane, an object that you can picture as a block of rubber, the stored energy grows in proportion to the volume. Membranes like these are central to the cyclic model of the universe.

String theorists use the shortened term *brane* to refer to all the types of membrane-like objects that have to be added to make string theory consistent. Then they add a prefix to specify the number of dimensions spanned by the membrane. For example, the term *2-brane* is short for "two-dimensional membrane," like a sheet. A string itself is a 1-brane. Pointlike entities are referred to as 0-branes. And there are branes with three or more dimensions. String theorists use the term p-branes (an awful pun) to describe an entity with *p* dimensions.

Even though theorists were forced to add more objects to string theory, this did not alter the uniqueness of its mathematical structure. In order to maintain mathematical consistency, the tension of every type of brane is strictly related to the string tension. Likewise, the rules that control how branes can interact with one another are unique and geometrical in character. For example, a 1-brane (string) can have its ends attached to a 2-brane, a 2-brane can roll up into a tube whose ends are embedded in other 2-branes, and so forth.

As theorists learned about branes and how to tie them together, string theory suddenly blossomed into a geometrical garden. Trying to understand strings without taking account of the other branes had been like attempting to comprehend an ecosystem by studying only one plant. With the discovery of all the other types of branes, new hopes arose for developing string theory into a realistic model of elementary particles.

In 1995, Witten took a giant step toward turning this dream into reality. He showed that all five known string theories are actually mathematical reformulations of a single underlying theory. The relationship between the five versions had been hidden because theorists had been considering extreme limits where strings are the only types of branes that affect physical phenomena. Being unaware of the other branes at the time, string theorists had thought that they were studying five unrelated string models, each with distinct properties. Like the old fable of the blind men and the elephant, they did not realize that they were describing extremities that were all parts of the same beast.

Furthermore, as Witten and Petr Hořava showed several months later, string theorists had missed the most interesting possibility: a sixth regime based solely on branes with two or more dimensions and without any strings at all. The Hořava-Witten model uses *ten* dimensions of space instead of nine and, in many ways, comes close to realizing Einstein's dream of expressing all fundamental physics in a unique and purely geometrical way. The new overarching perspective was dubbed *M theory,* a deliberately enigmatic name for the "master" or "mystery" theory, which theorists have been struggling to understand and to develop into a detailed theory of elementary particles ever since.

With the M theory breakthrough, the time had come to reconsider the relationship between fundamental physics and cosmology. Little by little, string theorists and cosmologists began to engage in discussions as to how the new geometrical picture of space and the fundamental forces might affect theories of the origin and evolution of the universe.

Can M Theory Inflate the Universe?

The first question cosmologists wanted to answer was whether M theory could fit, or even improve, their favorite cosmological model, the inflationary picture. Inflation had originally been inspired by grand unified theories and the particle-field picture. However, in the intervening years, the hope for grand unified theories had faded, the Higgs field was no longer considered a viable candidate for inflationary energy, and the search for alternative sources for inflationary energy had led to increasingly arcane ideas.

At first sight, M theory seems to offer a plethora of new candidates for this inflationary energy source. At every point along the usual three dimensions of space along which all of us move in our everyday lives, there are additional hidden dimensions wrapped in some complicated manner into a microscopic imperceptible volume. At present, the extra dimensions must be arranged in a low-energy state; otherwise their energy would increase the total energy density of the universe and cause the expansion of the universe to accelerate at a rate inconsistent with what is actually observed. Also, the size and shape of the extra dimensions must be fixed now, because any change would cause the masses and interactions of elementary particles to vary in ways that are not observed experimentally.

But when the universe first emerged from the big bang, the size and shape of the extra dimensions could have been highly distorted compared to today, arranged in a configuration with very high energy density. As the universe cooled, the configuration could have conceivably relaxed to the low-energy state found today by following an energy curve similar to those depicted for the Higgs field in the last chapter. Even if some dimensions relaxed quickly, the energy of one slowly relaxing dimension is enough to drive a period of rapidly accelerating expansion and enable the inflationary model to work.

By including branes, M theory presented even more opportunities for inflation to arise. Branes can be hidden like a magician's handkerchief, wrapped around the extra dimensions at every point in space. The energy stored in the branes can also act as inflationary energy. Also, for every brane, there is an antibrane partner that may also have been produced at the big bang. A colliding brane-antibrane pair can slowly attract each other and then annihilate, just like pairs of matter and antimatter particles, dumping their energy into matter and radiation and providing a natural mechanism for ending inflation.

As the number of possibilities became clear in the late 1990s, inflationary theorists became very optimistic. How could inflation possibly fail given this cornucopia of opportunity? Unfortunately, it turned out to fail very easily. The extra dimensions and branes have many ways of contracting and expanding and changing shape. If they are in a high-energy state when the universe comes out of the big bang, they can escape to a low-energy state by many routes. Typically, they take the fastest escape route, and that is so fast that inflation ends before the universe is smooth and flat. Holding together a high-energy configuration of extra dimensions for an extended period of time is like trying to hold together a structure made from slippery elastic bands and balloons. Most arrangements of the structure quickly fall apart.

Many theorists remain hopeful despite the failures. There are a very, very large number of possibilities to consider. Some estimate that there are 10^{1000} ways to arrange the extra dimensions, with branes and fields threaded through them. Some say the number is infinite. String theorists describe the different versions as a *landscape* of possibilities, which they are now combing to find a viable example of inflationary energy that leads to a universe with the types of elementary particles seen today. So far, the search has not been successful. And even if theorists do succeed in finding an example somewhere in the landscape of possibilities where the inflationary model seems to work, they will then have to explain why the universe chose that possibility, out of the huge number of alternatives. This is a critical and controversial problem that may ultimately be the Achilles' heel of the inflationary model or M theory or both. This issue is addressed at length in chapter 10.

It is not clear yet why it is proving so difficult to merge M theory and the inflationary model. Perhaps there is a clever idea to be discovered that will suddenly turn the current morass into a beautiful resolution. Then again, perhaps one should not be trying to take an old idea based on quantum field theory and three spatial dimensions and make it fit the completely different M theory picture. Maybe this calls for a radical alternative that fits more naturally with the revised laws of physics.

These thoughts were in the backs of our minds in the summer of 1999, when the two of us, together with our Russian colleague Valery Rubakov of the Institute for Nuclear Research in Moscow, decided to organize an international workshop at the Isaac Newton Institute in Cambridge. This was the fateful meeting described in chapter 1 that started us down the road to the cyclic model.

A New Start with Braneworlds

The plan for the workshop at the Newton Institute was to bring together as many leading string theorists and cosmologists as possible, with the hope of stimulating progress in connecting M theory with real-world cosmology. The political backdrop was distinctly unpromising. Just before the workshop, NATO entered the war in Kosovo, creating potentially dangerous divisions across Europe. Because the workshop was funded in part through a NATO-sponsored scheme, several scientists pulled out at the last minute. After some debate, the decision was made to go ahead. There is a strong school of cosmology in Eastern Europe, but there are few opportunities for personal interaction with cosmologists from the rest of the world. If the workshop were canceled, all the participants would have missed a rare chance to exchange ideas and build better links.

We asked Burt Ovrut, a highly respected string theorist and long-standing colleague of Paul's from his years at the University of Pennsylvania, to deliver a set of lectures on M theory that would be accessible to cosmologists. Ovrut is a leading expert in some of the most mathematically formidable aspects of string and M theory. On this occasion, he stepped back from the technicalities and presented an inspiring overview of the theory's essential elements.

He began by reminding the audience that M theory unifies the five versions of string theory by showing they are all just different limits of a single master theory. Furthermore, the master theory includes a sixth regime involving branes but no strings. Then, for the remainder of his talk, he concentrated on this sixth regime, the remarkable new geometrical picture discovered by Hořava and Witten. He asserted that this version deserves special attention because it offers new insights into fundamental physics and may provide the most direct way to relate M theory to laboratory experiments.

Ovrut is one of those physicists with artistic talent, so he was able to draw evocative pictures to illustrate Hořava and Witten's theory. He began with a drawing of two closely spaced parallel sheets. Although the sheets appeared two-dimensional, Ovrut asked the audience to imagine that the sheets were really nine-dimensional. (Time is an additional dimension, but, for simplicity, we will restrict ourselves to counting only the dimensions of space here.) The gap between the two sheets lies along the tenth dimension, the extra dimension that distinguishes Hořava-Witten theory from the other five string theories.

The sheets, Ovrut explained, are branes, but not ordinary ones. Branes normally have space on either side. The branes that Ovrut drew, though, are the boundaries of the extra dimension so that space only exists between the branes, not outside the gap. To emphasize his point, Ovrut called them "end-of-the-world" branes. As an analogy, consider a double-glazed window. The window itself consists only of two panes of glass and the air space between them. The two panes are like the branes and the gap between them is like the extra spatial dimension.

Ovrut then described a simplification of the Hořava-Witten picture that he and his collaborators had been exploring. Although the sheets are nine-dimensional, six of those dimensions are expected to curl into a ball so tiny that, on the scale of his drawing, they could not be discerned. The six curled-up dimensions are important in determining the properties of matter particles and forces. But since they would play no active role in his lectures, he said it was safe to ignore them. Now each sheet has only three (uncurled) dimensions, just like the three dimensions (height, length, and width) experienced in everyday life, and there remains only one extra dimension, the dimension between the two sheets.

From this sketch emerges a startling new vision of the cosmos: The observable universe lies on one of the branes, often called our "braneworld." It is separated by a tiny gap, perhaps 10^{-30} centimeters across, from a second "hidden braneworld." According to the equations of M theory, Ovrut explained, all the particles and forces we are familiar with, and even light itself, are confined to our braneworld. We are stuck like flies on flypaper, and can never reach across the gap to the "hidden" world, which contains a second set of particles and forces with different properties from those in our braneworld.

Although we cannot touch or see anything on the hidden braneworld, its effect can be felt because gravity exists throughout space and *can* cross the gap between braneworlds. For example, a lump of matter in the hidden braneworld generates a gravitational force pulling matter in our braneworld toward the point directly opposite the lump.

The gap is microscopic, perhaps 10^{-30} centimeters, so that the gravitational force would be strong even though the lump is invisible to us. One appealing possibility (though it is not required by the cyclic model) is that the dark matter accounting for most of the mass in galaxies and galaxy clusters consists of matter lying on the hidden braneworld.

Now picture a tiny bubble of membrane floating in the space between the two braneworlds, like a soap bubble floating between the two panes of glass in a double-glazed window. As first argued by Paul Townsend at Cambridge University and Michael Duff at Imperial College, London, the bubble represents a quantum of the gravitational field, a graviton, which can move back and forth across the gap. If it strikes one of the braneworlds, the membrane can become stuck to it, just as a soap bubble sticks to a glass surface. Hořava and Witten showed that membrane bubbles stuck on the branes in this way represent the particles and forces on the braneworlds, with different physical properties depending on how they spin and vibrate.

Ovrut, along with two postdoctoral fellows at the University of Pennsylvania, André Lukas and Daniel Waldram, and Kellogg Stelle of Imperial College, London, had gone a long way toward turning the braneworld picture into an impressively realistic model of particle physics. They showed that, with an appropriate arrangement of the six curled-up dimensions, the matter and force-carrier particles bound to our braneworld have physical properties closely matching those of the real-world particles in the table on page 72. They dubbed the resulting theory *heterotic M theory* since it replicated many features of the heterotic string model introduced in 1985 by David Gross, Jeffrey Harvey, Emil Martinec, and Ryan Rohm.

To see how heterotic M theory—based on ten dimensions of space and two braneworlds, with bubble-like membranes between them—can replicate the behavior of strings traveling in only nine dimensions of space, consider what happens to a soap bubble in the gap of a double-glazed window when the two panes are pushed together. At first, the bubble makes contact with only one pane. Then, when the second pane is close enough, the bubble makes contact and transforms into a tube connecting the two panes. Finally, when the panes are brought very close and the gap between them disappears, the tube turns into a thin ring. In a similar way, heterotic M theory can be transformed into a string theory by closing the gap between the two braneworlds. The membranes attached to each individual braneworld turn into tubes connecting them both and, finally, into a closed loop of string when the extra dimension becomes very small. Conversely, prying the braneworlds apart turns strings into tubes and, eventually, the tubes snap into membrane bubbles attached to one or other of the two braneworlds.

Ovrut next explained yet another reason to be excited by the M theory picture: it automatically incorporates supergravity, the theory which, as we mentioned in chapter 5, Stephen Hawking had

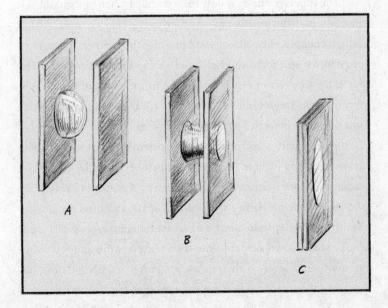

highlighted in his inaugural lecture. Supergravity had been popular in the early 1980s when theorists were still trying to construct unified theories based on fields and particles. It is a theory of gravity built around a powerful symmetry known as supersymmetry. A symmetry is a transformation of a system that produces no discernible change. For example, if you rotate a square by ninety degrees, it looks the same as when you started. We already introduced the notion of symmetry in grand unified theories when we described how different types of elementary particles—electrons, neutrinos, quarks, and so on—become indistinguishable at very high temperatures. Likewise, the strong, weak, and electromagnetic forces become indistinguishable (or symmetrical) facets of a single "grand unified" force at high temperatures. The idea of supersymmetry (or SUSY for short) goes much further than this. Instead of just relating particles to particles and force carriers to force carriers, supersymmetry says that, for every matter particle, there is a partner force-carrier and vice versa.

The supersymmetry is only fully revealed when one smashes elementary particles together at very high energy. For example, at high enough energies, the collisions will produce the partners of the matter particles and force carriers listed in the table on page 72. The search for supersymmetry is one of the prime motivations for constructing the Large Hadron Collider at CERN in Geneva, Switzerland, the most powerful particle collider ever built.

Supergravity also incorporates Einstein's theory of relativity and extends it to include the symmetry between particles and forces, leading to more mathematically consistent theories of elementary particles. Einstein's theory showed that space and time are neither absolute nor independent but, rather, are different aspects of a combined entity known as space-time. Two observers moving relative to one another disagree on their measurements of length and time, but Einstein's theory is based on a symmetry that transforms one set of measurements into the other. Supersymmetry goes further: in addition to exchanging measurements of space and time, it simultaneously exchanges force carriers with their partner matter particles and vice versa. When theories of quantum fields and particles incorporate supersymmetry, this turns out to dramatically improve their mathematical properties. Many (but not all) of the infinities which are otherwise present get automatically cancelled. And many of the parameters which one is otherwise free to adjust, making the theories arbitrary and adjustable, get fixed by supersymmetry to specific values. The more supersymmetry transformations that exist, the more unique and predictive is the theory.

In the 1970s, before string theory became fashionable, theorists had identified the simplest, most powerful version of supergravity with the greatest number of supersymmetry transformations. For a time, this theory was considered the leading candidate for the ultimate unified theory of nature. But with the rise of string theory in the mid-1980s, interest in supergravity declined. String theory included a more modest degree of supersymmetry. However, by replacing particles and fields with strings, string theory was more effective at removing the infinities which supergravity alone could not. Hořava and Witten's theory managed to combine the best features of both string theory and supergravity.

While Ovrut leaped ahead to outline the beautiful mathematics of supergravity and heterotic string theory, the two of us were riveted by his very first sketch of two parallel braneworlds. Ovrut had explained that the branes can move back and forth, so it is possible for the gap between them to open and close. We were both struck by the same thought: what would happen if two moving branes collided? It seemed likely that the collision would release a dense spray of energy, filling each brane with hot radiation. In other words, a brane collision might produce a big bang.

A Transdimensional Trip on the
West Anglia Great Northern (WAGN) Railway

Hardly able to contain our excitement, after the lecture we converged on Ovrut from different directions and cornered him to discuss the idea. Much to our surprise, he did not think it was crazy. Indeed, he and his collaborators had already begun to consider the implications of brane collisions for the properties of elementary particles, so the idea of using the collisions for cosmology seemed natural and enticing to him. All three of us wanted to discuss the idea right away, but that particular night a conference excursion had been arranged to see the play *Copenhagen* in London, and we had to leave immediately. Each taking a different route, the three of us arrived on the platform at the Cambridge station, bursting with ideas about how a braney big bang might work. As the WAGN train rumbled toward London, we brainstormed, launching a sequence of new ideas that would be the focus of our research efforts for the next two years.

We could see right away that brane collisions are unavoidable in M theory—nothing prevents two brane worlds from running into each other. Also, a brane collision would fill the branes with a nearly uniform density of matter and radiation. The details would be hard to calculate, but the physical picture was compelling. If M theory was right, the big bang just *had* to be a brane collision, we sensed. But if this was true, there would be huge implications.

First, the big bang would not be the beginning of time. If the big bang is a collision, then there must have been a time before the bang. In the mid-1960s, Stephen Hawking, building on methods developed by Roger Penrose in Cambridge, had shown mathematically how Einstein's theory of gravity breaks down as one attempts to trace time back to the big bang. As one follows time backward, all of space shrinks down to zero size and the density of matter and radiation increases to infinity. Einstein's equations then reduce to mathematical gibberish. This behavior had been known since Friedmann and Lemaître; what Hawking showed was that it was unavoidable and that the universe must have emerged from a singularity that could not be described by Einstein's theory.

What the brane-collision picture suggested was that M theory, with its improved description of gravity, might go further than just patching up the infinities of quantum gravity. It might actually be able to describe what happened at the big bang . . . and before it. For example, in the new picture, the two braneworlds do not shrink to zero size at the collision. Instead, they remain stretched out as they approach each other along the extra dimensions. Consequently, the density of the matter and radiation attached to each brane is finite at the bang, and the laws of physics, as amended by M theory, might still make sense.

Many obstacles had to be overcome to develop the primitive intuition into a realistic scenario for the universe. Each of the successes of inflation—explaining why the universe is uniform and flat and how it obtained the small density variations needed to seed galaxies—had to be replicated. Here, too, branes offered a fresh point of view. Rather than using inflation to smooth out the universe, one could appeal to symmetry. Imagine beginning in a state of perfect symmetry—supersymmetry, in fact—with empty, flat, parallel branes. Unlike panes of glass, branes are flexible. As the braneworlds are drawn toward one another, quantum jitter can cause them to wave around like bedsheets being rippled by a breeze. When two such rippled branes collide, their surfaces hit at different times in different places. The big bang would have occurred earlier at some points and later at others. Where the bang happened a bit earlier, there has been a bit more time since the big bang for the radiation and matter to spread out. These places would have a slightly lower temperature and density today. Similarly, the places where the bang happened later would have higher temperature and density today. They would be the places where the matter would self-gravitate and collapse to form galaxies and galaxy clusters.

But what started the branes moving toward one another in the first place? Ovrut was the expert on constructing realistic models based on heterotic M theory, and he explained that there are numerous physical effects that can create a springlike force between the two branes. Such a force might pull them together to cause a collision. The two of us pointed out that the quantitative nature of the force might also be important in generating ripples of the right type to explain the formation of galaxies.

Still, as exhilarated as we were by these ideas during the train ride to London, a great deal of work lay ahead before they could be turned into a viable model. Hořava and Witten had assumed a static, unchanging arrangement of branes, because this vastly simplified their analysis. The dynamic situation of colliding braneworlds was much harder to describe, and most of the theoretical tools and tricks that had been developed would be useless in this context. Nevertheless, we now had a tantalizing vision of a new cosmological scenario that went well beyond the inflationary model and employed the full power of M theory and branes to describe the big bang.

The Birth of the Ekpyrotic Universe

Theoretical physics is in some respects similar to certain Asian philosophies, according to which enlightenment is attained only at the price of great pain and personal suffering. In the case of theoretical physics, the suffering is mostly caused by the great complexity of the calculations necessary to develop and test theories. Einstein's theory of general relativity is already notorious in this respect. Conceptually, the theory is a model of elegance and economy. Its main equations can be written in terms of a few symbols that are easily memorized. But the only way to make predictions is to actually solve the equations. In general this is nightmarishly difficult, especially in physically realistic, time-varying situations like those Ovrut and the two of us were contemplating. Unfortunately, M theory was considerably worse. Not only were the equations more complicated, involving branes and gravity in higher dimensions, but they were only partially formulated. Nevertheless, driven on by hopes of a real alternative to inflation, the three of us started to sketch out techniques that could be used to describe two rippling braneworlds being pulled together toward a brane collision.

We agreed not to discuss these initial ideas with others yet: rather, we would develop them in private and publish the new cosmological model only if it showed signs of success. Otherwise, the negative response of other theorists to a sketchy idea could be so strong that the model might be buried before it had a chance to breathe. The criticisms could be anticipated. String theorists would say the work was premature and M theory was still too poorly understood to be applied to cosmology. And most cosmologists had already nailed their colors to inflation's mast: many would be sure to resist considering an alternative. To be accepted as a serious rival, the model would have to match each of inflation's successes. In particular, it would have to provide an alternative explanation for the nearly scale-invariant density fluctuations that seed the formation of galaxies.

Eighteen months of persistence followed. The first step was to have Paul's student Justin Khoury join the project. Justin had been studying particle physics, and he had recently decided that he wanted to work on a problem at the interface between string theory and cosmology. Paul and Justin talked almost daily. Communications with Ovrut occurred during biweekly trips between Philadelphia and Princeton and, at one point, during several weeks of intense research with Paul and Justin at summer workshop in Vancouver. With Neil, there were countless phone calls, e-mails, and faxes to and from Cambridge and, at times, South Africa, where Neil was busy helping to set up the new African Institute for Mathematical Sciences, a postgraduate center to support the advancement of science across the continent.

The picture we had in mind was one of two widely separated, parallel branes stretching to infinity in three directions. A tiny force existed between the two branes, causing them to attract and move very slowly toward each other along the fourth dimension over a long, perhaps infinite period of time. The force grew ever stronger as they approached, speeding their motion toward the collision. At the bang, the kinetic energy of the branes would be converted into hot radiation.

To achieve this condition, the attractive force had to be extremely weak when the branes were far apart but grow stronger as they approached. Surprisingly, the calculations showed that this type of force caused the branes to ripple in a very definite way. Minute quantum vibrations are intrinsic to all branes. These vibrations would be amplified by the attractive force. Where they brought the branes closer, the attraction would be larger. The branes would move faster and the collision would come sooner. After the collision, the radiation and matter would be stretched out by the expansion, so these places would end up with lower density. Places where the quantum vibrations pushed the branes apart would be the last to collide. The radiation and matter would undergo less stretching, so these places would have the highest density. They would form the seeds upon which galaxies and clusters of galaxies would later form.

The stunning finding was that, provided the force between the branes grows rapidly in strength as the branes approach, like that between two magnets or electrical charges, the density variations produced on the colliding branes were of just the right, nearly scale-invariant form needed to fit the real universe. Suddenly there was a genuine alternative to the inflationary model: a new solution to the conundrum of galaxy formation.

Inflation and the new mechanism are as different as can be. Inflation generates density variations in an infinitesimal fraction of a second, as the universe is undergoing a burst of super-rapid expansion. Assuming the brane collision occurs as described above, the new mechanism generates the variations over much longer periods of time as the two branes gently pull together. Yet the two models produce identical variations in the energy density across space. They make very different predictions about some other features of the universe, still to be measured (as described in chapter 9). But they agree on all measurements that have been made thus far.

As the four of us prepared to write our paper, we considered the question of what to call the new scenario. Friends frivolously suggested the Big Splat, the Brane Smash, and other humorous names. But, as many physicists do, we had a predilection for homage to the ancients. Paul approached classics scholars Joshua Katz from Princeton and Katharina Volk from Columbia University for advice. According to them, the new cosmological scenario sounded like the ancient Greek notion of *ekpyrosis,* in which the universe is born from out of fire. The word doesn't exactly roll off the tongue. Ovrut thought it sounded like some sort of skin disease. But we eventually settled for the *ekpyrotic universe,* and the name has stuck.

chapter seven

A Cyclic Model of the Universe

What we call the beginning is often the end
And to make an end is to make a beginning.
The end is where we start from.

—T. S. Eliot, *Four Quartets*

In late August 2001, when the ekpyrotic model was hot off the press, the two of us traveled to Rovaniemi, Finland, a town right on the Arctic Circle, to attend a meeting called COSMO-01. We had been invited to present our new interpretation of the big bang as a collision between branes, and the new mechanism for generating the energy density variations needed to make galaxies.

Whereas inflationary cosmology was over twenty years old and its premises had largely been accepted, the ekpyrotic model was brand-new and introduced several unfamiliar concepts simultaneously. Our original paper had been available on the Internet for some time, but it was long and technical and few would have had a chance to study it in detail. So when the organizers arranged a special discussion session following the main talks for us to compare the strengths and weaknesses of the inflationary and ekpyrotic models, it seemed like an excellent opportunity to address any criticisms or doubts. If the new ideas came through clearly, it would encourage others to join in the research.

Unfortunately, the forum quickly devolved into a raucous back-and-forth with a few committed enthusiasts of the inflationary picture, a snowstorm of silly barbs and minor quibbles. In the end, most of the audience probably left with the impression that the ekpyrotic picture was too exotic and confusing to be considered a serious contender. As the two of us left the auditorium, we were so frustrated by the lost opportunity that we agreed not to exchange a word about what had happened until breakfast the next morning.

Neither of us slept much, and even the delicious traditional Finnish breakfast fare couldn't improve our grumpy moods. As we ate, we reviewed the previous evening's discussion, turning eventually to the one valid point that had been raised. It was the same point we had been debating from the start: the assumption that the universe begins with two empty, perfectly flat, parallel branes.

Our first paper on the ekpyrotic model had appealed to a mathematical principle, symmetry, suggesting that the two branes might naturally start out in this state, gradually drifting toward each other over an infinite period of time before colliding and setting off the big bang. In the past, others had appealed to symmetry to explain the initial state of the universe. For example, about a decade earlier, Gabriele Veneziano, the same CERN theorist who founded string theory, had introduced a cosmological model called the pre–big bang model. Although his idea was phrased in terms of field theory rather than branes and extra dimensions, and differed in many other respects from the ekpyrotic picture, Veneziano was also endeavoring to make a cosmological model in which the universe existed for an infinite time before the big bang. He, too, had invoked symmetry to set simple conditions in the infinite past. But using symmetry to pick a smooth initial condition is less compelling than having a smoothing mechanism that can take a very wide range of initial conditions and make them flat and smooth. Indeed, the smoothing effect was one of the most attractive features of high-energy inflation. Of course, the goal of the ekpyrotic study was to determine if it was possible to formulate a cosmology without high-energy inflation.

As we continued the discussion during a walk along the Kemijoki River in the bright Arctic sunlight, one of us finally asked aloud the looming question that each had been formulating to himself: "Is it possible to find a physical smoothing mechanism for the ekpyrotic model that does not entail a period of inflation after the bang?"

Instantly, the conversation switched to rapid-fire brainstorming, which went something like this:

Could there be something that stretches and flattens the branes before they collide?

There is unlimited time before the bang, so the mechanism does not have to be so rapid or involve such high energies as inflation.

Actually, come to think about it, we know of something that could do the trick: dark energy.

If dark energy dominated the universe before the bang, then it would stretch the branes and make them flat, smooth, and parallel. The problem is that the dark energy observed today has come to dominate the universe only recently.

Hold it! Maybe we've missed something. When we introduced a springlike force that draws the branes together, it comes with a potential energy that could act just like dark energy. Only this seems like a step backward. One of our complaints about the inflationary model is that it requires two forms of unseen energy, one to drive inflation and one to act as dark energy. Now we would be introducing one form of dark energy to make the branes flat and smooth and one to account for today's cosmic acceleration. . . .

Wait a minute! What if the dark energy before the bang is the same as the dark energy today? What if the dark energy then and the dark energy today are both the springlike energy due to the interbrane force? The dark energy could dominate for a long time before the collision, causing the branes to become empty, flat, and smooth. Then, as the collision approaches, the potential energy is converted into brane kinetic energy . . .

. . . And at the collision itself, some of the brane kinetic energy is converted to matter and radiation. The branes bounce back to the separation they have today, so the springlike potential energy is restored.

But the density of matter and radiation created at the collision far exceeds the potential energy after the collision. The potential energy would only be noticeable again after nine billion years of expansion had passed and the density of matter and radiation fell below the potential energy.

Only then would the springlike potential energy take over again, just as it did before the bang. Once again, it would act like a source of dark energy that causes the stretching of the branes to accelerate, just what we are witnessing today. . . .

Of course, if it could happen once, there is nothing to stop the whole process from happening again, and again, and again. The bangs could continue forever.

Suddenly and inadvertently, we had revived an ancient idea that we had been taught was impossible: a cyclic universe.

Don't ask who said what in this conversation. Once the brainstorm started, the sentences shot back and forth too rapidly for us to keep track. In a few instants, the mood had changed from depression to elation. At the end of the walk, we said goodbye and Neil headed off to the airport. Only then did the realization of what had just happened begin to dawn on us. Here, finally, was a new kind of solution to the flatness and smoothness problem. Here also was a reason why dark energy ought to exist: not just an ingredient added to make the model fit the astronomical observations, but an essential and fully integrated element of the cosmological picture. And here was a way to turn the single collision between braneworlds into a remarkable kind of cyclic event, allowing a transformative view of cosmic history. Instead of leaving the fracas in Finland demoralized, the two of us flew off with sky-high hopes.

The New Cyclic Universe

All we had when we left Finland was a picturesque vision. Now the vision had to be translated into precise mathematical equations that could be checked for consistency with the known laws of physics and whose predictions could be compared to current observations and experiments.

The key steps were taken in the weeks following the Arctic brainstorm, in the wake of the 9/11 tragedy. Paul flew back to Newark a few days earlier, but by the time Neil got to Princeton to work on the new idea, the world had changed. Neil was one of the first to fly from Europe into Newark, just after Newark Liberty National Airport reopened. Especially in Princeton—where friends and neighbors had died in the Twin Towers and others had perished on the airplanes—the tragedy was constantly on our minds. In a certain sense, it was therapeutic to escape from the horror and ponder times when the universe was pure and simple.

After a couple of very intense weeks of calculations, a detailed cyclic model began to emerge. The checks and cross-checks got more and more complicated, but the model itself remained remarkably simple. With one key assumption—that the branes attract with a springlike force that becomes weaker as the branes separate—everything worked, in principle.

Chapter 3 has already introduced the cyclic picture, but from the perspective of someone who is aware of only three spatial dimensions. With the concepts introduced in the chapters that followed, it is now possible to translate that description into a dynamic theory of extra dimensions and braneworlds. As described below, the cyclic universe can be built from two braneworlds drawn together by a springlike force and colliding at regular intervals.

The springlike force can be characterized by an "energy curve" showing how the energy stored depends on the separation between the branes. The shape of the curve bears some resemblance to the shape of the energy curve needed for the new inflationary model (page 97). But here the curve is shifted downward so that the minimum is at large negative energy density, and today's universe corresponds to a point on the plateau rather than the minimum.

For inflation, the curve describes how the vacuum density depends on the value of the inflation field, the field driving the inflationary expansion. (By convention, the inflationary energy curve is flipped left to right, as shown in previous illustrations.) The plateau is set *high*, at a level that is 10^{100} times greater than the dark energy density today, to drive a period of high-energy inflation right after the big bang. And the universe follows a one-way path down the curve. Once it reaches the bottom, it remains there.

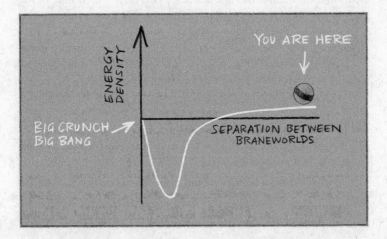

All of these aspects are different in the cyclic universe model. Here the energy density (shown on the vertical axis) is the stored energy associated with the springlike force between a pair of branes. The curve indicates how the stored energy depends on their separation. As in the case of inflation, the energy curve has a plateau, but here the plateau is set much *lower*, at a level equal to today's dark energy density. The idea is that the present universe corresponds to the point on the plateau indicated by the ball. The equation describing how the brane separation changes during a cycle is similar to the equation describing a ball rolling back and forth along a surface with the same shape as the energy curve. Since the plateau is very flat, the separation is nearly constant today, and the stored energy behaves just like dark energy, causing the expansion of the universe to speed up. The brane separation inevitably decreases, however, due to the slight downhill slope of the energy curve. Beginning from the present, the separation slowly shrinks. As the energy curve dips strongly to negative values, the branes speed up. When the separation hits zero, the branes collide in a big crunch, followed by a big bang as the branes separate once again. Then the branes fly apart, rapidly returning to their original separation on the plateau of the energy curve. Hence, in spite of the superficial resemblance of their energy curves, the inflationary and cyclic models predict completely different cosmic histories: a one-way path down the energy curve for the inflationary picture and a regularly repeating back-and-forth motion for the cyclic model.

Following a full circuit of the cyclic model makes it possible to appreciate how the different stages naturally interweave, using the natural transformation of one form of energy into another and yet another to keep the cycles going.

The description begins one full cycle ago, at the position in the centerfold labeled "You are here," and proceeds clockwise. A full cycle ago, the physical conditions were the same as they are today, and as they will be a cycle from now. Hence, the same position on the figure could also be labeled "You will be here a cycle from now" and "You were here a cycle ago." As noted above, the present brane separation, and, hence, the brane separation a cycle ago, corresponds to a point on the energy plateau. The evolution of the brane separation along the energy curve is the same as that of a ball rolling on a surface of the same shape. The slope of the plateau is very gentle at first, so the separation decreases very slowly. In the meantime, the universe continues to expand. Matter and radiation are diluted away, but the energy density stored in the springlike force, equal to the height of the plateau on the energy curve, acts just like dark energy and drives an accelerating stretching of the branes. That is, the separation between branes is nearly constant while the branes themselves stretch exponentially.

For the next trillion years, the branes double in size along each of their three dimensions a hundred-fold times. The stretching eradicates any wrinkles on the two branes, causing them to become almost perfectly flat, smooth, and parallel to one another. In this way, dark energy naturally restores the branes to a nearly pristine condition, readying them to begin a fresh cycle of evolution.

The dark energy does not dominate forever. The springlike force provides a natural shutoff valve. By slowly drawing the branes together, the force causes the universe to make its way slowly down the energy curve until, after about a trillion years, the potential energy density reaches zero. Without positive potential energy density to drive it, the accelerated expansion stops.

At this point, the universe enters a new era, in which the space between the branes keeps decreasing but the branes themselves stretch very, very slowly. This period is called the contraction phase, although it is important to note that only the extra dimension shrinks as the branes approach. As the branes get closer and the energy curve becomes steeper, the attractive force between branes

THE CYCLIC UNIVERSE

INTERBRANE FORCE DRAWS
BRANES TOGETHER, AMPLIFYING
QUANTUM WRINKLES.

A TRILLION YEARS
AFTER THE BANG:
BRANES ARE EMPTY,
FLAT, AND PARALLEL.

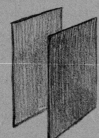

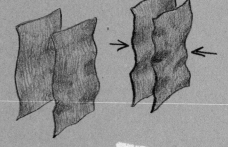

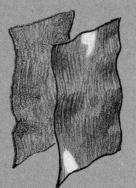

YOU ARE HERE →

DARK ENERGY TAKES OVER,
DRIVING ACCELERATED EXPANSION
THAT BEGINS TO SPREAD OUT
GALAXIES AND MATTER.

TWO BRANES ENGAGE IN AN ENDLESS CYCLE OF COLLISION, REBOUND, STRETCHING, AND COLLISION ONCE AGAIN

WRINKLED BRANES COLLIDE, CREATE SLIGHTLY NONUNIFORM HOT PLASMA, AND REBOUND.

A MICROSECOND AFTER THE BANG: BRANES REACH MAXIMUM SEPARATION BUT CONTINUE TO STRETCH RAPIDLY, FILLED WITH RADIATION

RADIATION DILUTES AWAY; MATTER DOMINATES AND CLUSTERS AROUND NONUNIFORMITIES TO FORM GALAXIES AND STARS.

grows in strength and some of the stored energy is converted into kinetic energy—the energy of motion. The mixture of potential and kinetic energy is the high-pressure form of dark energy described in chapter 3 that keeps the universe smooth and flat all the way up to the big crunch.

Because the branes are still stretching as the extra dimension contracts, the density of matter and radiation on the branes, which is diluted during the dark energy phase, remains diluted throughout the contraction phase. This is how the cyclic model avoids the usual problem of reaching infinite temperature and matter density at a big crunch.

Another important feature of the contraction phase is the gravity boost. If not for gravity, the braneworld "spring" would wind down, because with each new collision, some brane kinetic energy is irreversibly converted into matter and radiation. Eventually, the universe would settle at the minimum of the energy curve, just as in the inflationary model. But gravity saves the day by automatically converting some of its energy to brane kinetic energy. Including this boost effect, when the brane worlds collide, the branes have enough energy to bounce apart and return to the same place on the plateau where they began a cycle before. In this way, cycles can repeat at regular intervals.

In the contraction phase, quantum ripples on the branes are amplified in the manner described at the end of chapter 6, into a scale-invariant pattern. The amplification operates slowly, and on macroscopic scales. For example, the ripples that form about a millionth of a second before the big bang are a few meters across. These ripples are the ones that will ultimately seed the formation of galaxies. This part of the story is similar to the ekpyrotic model, and so it is referred to as the ekpyrotic mechanism. As a result of the wrinkles, the collision between braneworlds occurs at slightly different times in different places along the branes, causing some places to heat up and expand before others. The ensuing density variations thus account for the WMAP image and the seeds needed for galaxies.

The contraction of the extra dimension ends with the collision of the two branes, literally the next cosmic "big bang." In the collision, the two branes retain most of their kinetic energy so that they rebound at nearly the same speed with which they approached. However, as a result of their violent impact, some of their initial energy is converted into a hot spray of particles and radiation, filling each brane with a plasma at a temperature of 10^{23} degrees or so. According to Hořava-Witten theory, the two branes have opposite tensions: one positive and one negative. The negative tension brane has the remarkable property that dumping positive matter and radiation energy on it, for example a hot plasma, actually reduces its inertia and causes it to speed up. If the collision produces more plasma on the negative tension brane than on the positive one, then the branes actually recoil from the collision faster than they came in. This gain in speed is another vital element of the cyclic model, allowing the cycling to keep going even as more and more plasma energy is created at each new collision. A mere 10^{-25} seconds after the bang, the branes have come to rest at their original positions.

The conditions just after the bang are worth noting. The braneworlds have become nearly flat and parallel, and emptied of matter, radiation, and any other debris from the last cycle, during the period of dark energy domination and the ensuing contraction. The tiny quantum wrinkles on the branes have been amplified into just the right kind of temperature and energy fluctuations to explain the distribution of galaxies and the WMAP picture. At the same time, because the maximum temperature reached at the brane collision is much lower than that required to make magnetic monopoles, the monopoles never form and, hence, never cause the problems they did for the original big bang picture. In short, the cyclic universe has emerged from the bang with all the attributes that a period of high-energy inflation was supposed to provide. Hence, no high-energy inflation is needed!

The rest of the history up to the present is the period described as Act Two in chapter 2, and is the same as in the inflationary picture. The brane separation remains fixed through the radiation-dominated epoch. The branes stretch; radiation cools to form the first elements; massive particles take over the universe; the particles clump to form galaxies and more complex structures; and, 14 billion years later, the universe is back at the point marked "You are here" in the centerfold diagram. (This time, we really mean *today*.) The physical conditions of the universe today are the same as they were a cycle ago, or a cycle before that.

Calculations can be done using well-accepted methods for every stage of the cyclic model except one: the collision itself. To deal with the collision, one has to go beyond Einstein's theory of gravity and describe space and time in terms of more basic entities: strings and branes.

In 2004, with Malcolm Perry of Cambridge University, we analyzed a brane collision like the one that occurs at the transition from big crunch to big bang in the cyclic model. A calculation including all the components and features of M theory is too complicated for theorists to handle at this point in time, so we studied a simplified version that focused on what could be the Achilles' heel of the cyclic picture: the quantum production of matter and radiation in the last instants before the collision. First, there is the possibility that the energy of a created particle becomes infinite as the separation between the branes shrinks to zero. Second, even if the energy of individual particles remains finite, there is the possibility that their combined energy density becomes infinite. Either effect could cause the over-production of black holes and prevent the universe from cycling. When the branes are very close, all the quantum particles are described by membranes wrapped into tubes connecting the two branes. As the separation between branes decreases, the tubes become short and squat and behave more and more like strings, as the soap bubble picture on page 141 suggests. The various oscillations of these strings correspond to all the matter particles and force-carrier particles in the theory, including the graviton.

To our delight, the calculations showed that membrane tubes evolve smoothly in time right up to and through the brane collision, revealing no obstacle to surviving the transition from big crunch to big bang. We also studied quantum effects that cause new membrane tubes to be created at the bounce so that, as the branes move apart and the tubes snap, a finite density of hot particles and radiation is produced on each of the two branes, just as envisioned in the cyclic model. Although the model is simplified, these results strongly indicate that string and M theory are capable of describing a realistic transition from a big crunch to a big bang, something Einstein's theory of gravity could never do.

Nevertheless, more work is needed to prove that the quantized, vibrating membrane and string states pass smoothly through the bounce. To be sure, many string theorists remain skeptical about the ability of the crunch to transform into a bang, and this makes them uneasy about the whole cyclic concept. There is no way to resolve the issue other than performing detailed calculations to see if they result in any infinities or ambiguous answers. Today, many theorists around the world are hotly pursuing the problem.

Key Elements of the Cyclic Model

As described here, the cyclic model is built on several key elements. First, from string theory and M theory come the ideas of branes and extra dimensions, which allow for a big bang where the density of matter and radiation is finite. Second, observations of the present universe indicate the existence of a form of energy, dark energy, that is ideal for smoothing and flattening the universe. The same dark energy acting before the big bang could explain why the universe is smooth and flat on large scales today. Finally, the decay of the dark energy leads to a buildup of energy sufficient to power the big bang, while simultaneously generating density variations that can give rise to galaxies after the bang.

The model described in this chapter is the simplest, and the one most directly connected to string theory and M theory. But there are many other theoretical frameworks in which a cyclic model might be viable. For example, in the simplest version of Hořava and Witten's model, the gap between the two braneworlds is a mere 10^{-28} centimeters at the present time. This distance is set by matching the strength of gravity and that of the other forces in particle physics. More complicated arrangements of the extra dimensions and branes are also possible. A few years before the cyclic model was invented, theorists had started to play with some of these options. Savas Dimopoulos at Stanford University, Nima Arkani-Hamed at Harvard University, and Gia Dvali at New York University considered theories in which the extra dimensions of space could be as large as a millimeter. Lisa Randall at Harvard and Raman Sundrum at Johns Hopkins University considered a further generalization where the distance between braneworlds could be arbitrarily large. Their papers initiated an explosion of braneworld ideas. Although the cyclic model was originally inspired by the Hořava-Witten framework and the link to M theory, any one of these interesting generalizations may also be compatible with the cyclic picture.

In the cyclic universe model described thus far, the cycles continue regularly forever into the past and future. At present, there is no known theorem or principle that prevents this from occurring. Some theorists have argued that the number of cycles must be finite; otherwise, the universe would be like a perpetual-motion machine. All of us were taught in school that perpetual-motion machines are impossible because they violate either the conservation of energy or the second law of thermodynamics, the law that entropy (or, equivalently, disorder) always increases, a basic principle of physics to be discussed in chapter 8. These two fundamental principles have been a major roadblock for all cyclic models of the past.

However, as we have explained, in the cyclic model there is a gravity boost that converts gravitational energy into the energy of motion of the branes, so that energy is still conserved in each and every cycle. A unique property of gravity that has been well known since Newton is that there is no limit, in principle, to how much energy you can borrow from it. As it turns out, braneworlds with springlike forces have the ideal properties needed to set up an automatic borrowing procedure that converts gravitational energy to matter and radiation once every trillion years. There is no known reason why this borrowing could not continue forever or why a "beginning" is required.

Although the cyclic model does not require a beginning of time, it is compatible with having one. One could imagine the sudden creation from nothing of two infinitesimal spherical branes arranged like two concentric soap bubbles, both of which undergo continuous expansion as well as regular collisions with each other under the influence of an interbrane force. Both brane bubbles would grow enormously with every new cosmic cycle. After several cycles of expansion, the pair of branes would appear very flat and very parallel to any observer like us, with access to only a limited region of space. For such an observer, there would be little difference between this universe with a beginning, and a universe in which two flat, parallel branes had been colliding forever into the past.

The final chapter of this book will discuss the possibility that the universe slowly alters its state from one cycle to the next. In the simple version of the cyclic model explained so far, all the physical properties of the universe are the same, on average, from cycle to cycle. The Earth itself exists only in this cycle, but Earth-like planets would presumably exist in every cycle and so, perhaps, would life. However, switching from the inflationary picture with an age of 14 billion years to a cyclic picture in which the universe is much older should open one's mind to the possibility of evolutionary time scales that are far greater than ever before imagined. Then some properties thought to be constants, like the masses of elementary particles, the strengths of the various forces, and the cosmological constant, could actually vary over very long periods. This notion holds some promise of resolving some of the hardest problems of cosmology and fundamental physics, and is perhaps one of the most important new ideas to emerge from the cyclic picture.

chapter eight

The Last Question

The law that entropy always increases—the second law of thermodynamics—holds, I think, the supreme position among the laws of Nature. If someone points out to you that your pet theory of the universe is in disagreement with Maxwell's equations—then so much the worse for Maxwell's equations. If it is found to be contradicted by observation—well, these experimentalists do bungle things sometimes. But if your theory is found to be against the second law of thermodynamics, I can give you no hope; there is nothing for it but to collapse in deepest humiliation.

—Sir Arthur Stanley Eddington, *The Nature of the Physical World*

Neither of us were fully aware of the long history of cyclic cosmologies when our first paper on the subject was published, in 2003; nor was it our original intent to construct a theory of a regularly repeating universe. We had been drawn to the cyclic universe by a roundabout line of reasoning. We began from the simple picture suggested by M theory of a universe composed of two branes separated along an extra dimension. This suggested the possibility that the big bang might be the result of a collision between two branes. Exploring this idea led to a new explanation of many of the detailed features of the universe, including the distribution of galaxies and the temperature variations of the cosmic background radiation. But if the big bang could occur once, it could occur many times. As we'd realized that fateful morning in Finland, the recently discovered dark energy is just what is needed to restore the branes to a flat, parallel state, thereby allowing the collisions to repeat in a regular manner.

We had not anticipated that creating a cosmological theory founded on the most recent breakthroughs in string theory would revive a long-standing debate dating back to ancient times. Investigating the connection to earlier ideas was certainly fascinating from a historical point of view. More than this, though, it was important to check whether any of the scientific criticisms that had been hurled at cyclic models in the past, and had caused them to be largely ignored for over fifty years, also applied to the new cyclic model.

Virtually every cosmological model throughout recorded history falls into one of three categories. The first might be labeled "created universe," since it assumes that the universe sprang into being a finite time ago and has been steadily evolving since. A second category, "unchanging universe," incorporates models in which the universe remains the same forever. And the third is "episodic universe"; it contains cosmologies with repeating epochs of creation, evolution, and dissolution, where the repetitions may be regular or irregular. At most periods in history, ideas from all three categories were being discussed and debated in one place or another around the world.

In fact, the period since the discovery of the cosmic background radiation in 1964 has been exceptional. Only one view, belonging to the "created universe" category, has been widely held. Curiously, the founders of the big bang picture—Friedmann, Lemaître, Einstein, Willem de Sitter, and Gamow—were uncertain whether the big bang model belongs in this category. There was a considerable diversity of opinion about what the model implies about the beginning of space and time, and a persistent interest in the cyclic option. The founders of modern cosmology never reached a conclusion, and the issue was not settled by their successors. Instead, following the discovery of the cosmic background radiation in the 1960s, the focus of research switched from the issue of origins to empirically testable matters, like the formation of the first nuclei, atoms, stars, and galaxies. Since then, most cosmologists have simply taken for granted that the big bang is the beginning of space and time. That is how the big bang is explained both in technical textbooks and in popular descriptions. Now the cyclic model provides a scientific reason to reassess that assumption.

The First Cyclic Models

Earlier thinkers relied on a mixture of arguments, based on common sense, everyday experiences, philosophy, and even divine inspiration. The variety of opinion is striking, and some of the ancient views seem, with hindsight, remarkably modern.

Common experience no doubt stimulated many of the age-old ideas. The notion of a created universe was probably modeled on human conception and birth. Similarly, a cyclic universe naturally springs to mind when one considers the rhythmic variations that dominate all of our lives: the daily cycle of sunrise and sunset and the annual cycle of the seasons.

Ancient Hindu cosmology presents a remarkably detailed and quantitative vision of cyclic evolution. The full picture has cycles within cycles within cycles, where each level of cycle has a different duration. The levels correspond to various timescales in the lifespan of Brahma, the god of creation. For example, one kind of cycle corresponds to a day and a night in Brahma's life, another to a year, yet another to one hundred years, and so on. Converted to Earth years, some of these levels have cycling periods that are surprisingly similar to timescales of interest in contemporary cosmology. A day and a night in the life of Brahma lasts a *kalpa,* a period of 8.64 billion years, roughly the duration of the matter-dominated epoch in modern cosmology, during which galaxies formed. The Vishnu Puranas say that each of these cycles is followed by a drought that lasts until all the waters dry up. In modern cosmology, matter domination is followed by dark energy domination, during which no new galaxies form. The next level of cycle, corresponding to a year of Brahma or 360 *kalpas,* lasts 3.11 trillion years. This duration is roughly that of a single cycle in the cyclic model. The lifetime of Brahma is a factor of one hundred longer, after which the universe takes a respite. Exactly what happens next is less clear. There have been numerous Brahmas before and there will be numerous ones to follow. The Hindu texts do not say how long the interval between Brahmas is. The evolution may be cyclic or it may be sporadic.

In the West, cyclical cosmology was a dominant idea for nearly six centuries, beginning around 500 B.C. The concept can be traced back to a disagreement between two of the earliest renowned Greek philosophers, Parmenides of Elea and Heraclitus of Ephesus. Parmenides' view, which Plato greatly admired, was that ideas are real and sensations are illusory. If thought is reality, then anything one can conceive must exist. Parmenides reasoned that since you cannot conceive of something "not existing" without first thinking of the thing itself (which means it already exists), then it is logically impossible for existence to have a beginning or an end. Hence, he concluded, everything endures and nothing changes.

Heraclitus held the opposite point of view. "All is flux" was his dictum, meaning that everything changes and nothing endures. The goal of philosophers, he argued, should be to understand how things change. Today, the Heraclitean point of view is recognized as underlying modern science, economics, politics, and history.

Following Heraclitus, the Stoics introduced the concept of ekpyrosis, from the word meaning "out of fire." As discussed in chapter 6, we'd adapted this ancient term as the name of the predecessor of the cyclic model, in which only one brane collision was considered. The Greek notion was that the universe begins and ends in a giant conflagration, with a period of normal evolution in between. The concept had many variants. There could be one conflagration, or an infinite number of them. They could be irregular and sporadic, or regular and periodic. The cycles could be exact repeats of history or only statistically similar.

In the periodic version, Cicero called the cycle the *annus magnus,* or "great year," and it was calculated to have a duration of ten to twenty thousand years. In his treatise *On the Nature of the Gods* (Book II, chapter 46), Cicero explained, "There will ultimately occur a conflagration of the whole world . . . nothing will remain but fire, by which, as a living being and a god, once again a new world may be created and the ordered universe restored as before."

The cyclic view became less popular in Europe as Christianity took hold, due in large part to the interpretation of the Book of Genesis by early Christian theologians. Many today read translations of the opening lines and interpret them as describing a universe that is created from nothing in a singular event. In the original Hebrew, though, the meaning is ambiguous at best. Rabbinic scholars are not sure whether "heaven and earth" refers to the entire universe or just the Earth. And a common interpretation is that this creation was from preexisting material, meaning that space, time, matter, and energy existed before the moment of creation. Some rabbinic interpretations even envisage that worlds have been created repeatedly, with earlier versions having been destroyed because their Maker found them unsatisfactory. In truth, though, these cosmological issues have never been considered central to Jewish religious thinking.

In Christianity, scholars have tended to link the origin of the universe more closely to the foundations of the religion, which is perhaps why there has been greater tension between science and this faith, compared to other religions. For example, Saint Augustine argued strongly for a created universe, ascribing it an age of five thousand years. And he rejected any interpretations of biblical passages that suggest a more enduring existence. In his *Confessions* (Book XI, chapter 12), he wrote of the first lines of Genesis, "For the man who says, 'What did God do before He made heaven and earth?' I do not give the answer that someone is said to have given in jest: 'He was preparing hell, just for those who pry too deeply.' It is one thing to see the answer; it is another to laugh at the questioner—and for myself I do not answer these things thus. More willingly would I have answered, 'I do not know what I do not know,' than cause one who asked a deep question to be ridiculed—and by such tactics gain praise for a worthless answer. Rather, I say that thou, our God, art the Creator of every creature. And if in the term 'heaven and earth' every creature is included, I make bold to say further: 'Before God made heaven and earth, he did not make anything at all.' "

Although many Christian theologians adopted Saint Augustine's view, the concept of a cyclic universe retained some popularity. By the nineteenth century, the idea began to appear more frequently, even among popular writers. Edgar Allan Poe wrote an essay entitled

"Heureka" in which he proposed a specific "model" of the universe that is remarkably like the ancient ekpyrotic picture. In his vision, the universe results from the explosion of some simple, uniform state of matter in "one instantaneous flash." These initial atoms are blown apart by the expansion but then are drawn together by an attractive force that causes them to collapse into a uniform initial state again, from which a new explosion arises.

The German philosopher Friedrich Nietzsche also advocated a repeating universe, but his arguments in support of the idea were different from those of the early Greek philosophers. He thought that since there is no end to time, and presumably only a finite number of possible events and things, everything now existing must recur. Nietzsche's model of "eternal recurrence" was popular in the late nineteenth century and perhaps even stimulated the thinking of mathematical physicists like Ludwig Boltzmann and Henri Poincaré in their studies of heat and dynamical systems. Boltzmann's and Poincaré's ideas continue to influence modern cosmology.

Cyclical Models and the Birth of Modern Cosmology

Einstein and the highly literate band of physicists trying to construct a cosmology based on general relativity were familiar with the historic ideas. Einstein's initial preference was a static universe because, like Parmenides, he could not imagine that the universe has a beginning. But once Hubble discovered the expansion of the universe in 1929, the static model was eliminated.

Many conventional histories suggest that cosmologists turned next to the big bang idea, that the universe had a moment of creation and has been expanding and evolving ever since. The true history is more textured, though. If anything, Einstein and the other leading cosmologists of his day were reluctant to accept the concept of a "beginning." By the 1930s, they understood from Hubble's discovery that Friedmann and Lemaître's description of an expanding universe could describe the evolution of the universe over many billions of years. But they could envision more than one possibility for what happened at the bang itself, including the possibility that there was no beginning. Their reasoning has been largely forgotten, but is interesting to recall in light of the revival of the cyclic model.

Actually, cosmologists' doubts about the "beginning" predate Hubble's discovery. Although Alexander Friedmann wrote down the key equation underlying the big bang picture in his seminal paper of 1922, he himself was ambivalent about the issue. A mathematical physicist by training, Friedmann was interested in finding new cosmological solutions to Einstein's equations of general relativity. Einstein had published his static model of the universe five years earlier. The Dutch cosmologist Willem de Sitter soon followed with a model with a cosmological constant and no matter or radiation and showed that the universe can expand or contract. Friedmann wanted to show that there is yet another, more realistic type of model: an expanding universe with matter and no cosmological constant.

In fact, Friedmann identified not one but three possible models: the open, flat, or closed models described in chapter 2. He showed that the first two both have a big bang singularity, and then expand forever. The matter is gravitationally self-attractive, but there is insufficient mass density in these models to overcome the expansion of the universe, so the expansion never stops. He called this possibility a "monotonic world." (Mathematicians use the term "monotonic" to refer to a quantity whose value always increases or always decreases.)

Friedmann also studied the third, closed model, which he termed a "periodic world." In this case, the mass density is so high that the gravitational attraction drawing matter together is sufficient to halt the expansion of the universe and cause it to recollapse in a big crunch. Friedmann's classification indicated that he thought time could "vary from minus infinity to plus infinity, and then we come to a real [endless] periodicity."

Lemaître's background was very different from Friedmann's. Lemaître received his scientific training in astronomy and physics only after studying theology and becoming ordained as a priest. Five years after Friedmann, he rediscovered for himself Friedmann's cosmological solutions, and developed a detailed cosmological model incorporating the modern description of radiation, matter, and the red shifts of galaxies. (He did not learn of Friedmann's work until after he published his own paper.) Because Lemaître's work painted a more complete picture of the physical conditions in the early universe, he is often referred to as the "father of the big bang." As a physicist, he was well aware that his equations imply the density and temperature reached infinite values in the past, and that this might suggest some kind of beginning, but he was not convinced that the equations could be trusted to describe the conditions at those early times. In fact, he saw this "beginning" as a flaw in his model and struggled with it throughout the rest of his career.

One may have guessed that, as a Catholic priest, Lemaître would embrace the "beginning" required by his cosmological model as a scientific manifestation of the creation story presented in Genesis. Lemaître, though, was resistant to the idea of mixing religion and science: "The researcher makes an abstraction of his faith in his researches," he wrote. "He does this not because his faith could involve him in difficulties, but because it has nothing directly in common with his scientific activity. After all, a Christian does not act differently from any non-believer as far as walking, or running, or swimming is concerned."

In his thesis work, Lemaître tried to evade the "beginning" by positing that, before the universe began to expand, it existed for an infinite amount of time in a *nearly* static state, very similar to Einstein's model. His notion was that the expansion would begin because the static condition is not perfectly stable. The universe would ease into expansion after spending a semieternity in Einstein's static state. In this way, he hoped to get the best of both models, an expanding universe that exists forever. By 1931, he'd determined that this melding of the two models could not be achieved using Einstein's equations of general relativity, and so he appealed to quantum physics for a solution, arguing that the universe began in a nearly static, pure quantum state that he called a "primeval atom."

Einstein's opinions on these matters make a particularly interesting case study. By 1931, Edwin Hubble and Milton Humason at the California Institute of Technology had extended Hubble's earlier analysis to more distant galaxies and solidified the case for an expanding universe. Einstein was now convinced his static model of the universe was wrong and prepared to write a paper stating that his cosmological constant should be abandoned. "If Hubble's expansion had been discovered at the time of the creation of the general theory of relativity, the cosmologic member [his term for the cosmological constant] would never have been introduced since its introduction loses its sole original justification," he explained in a retrospective that he appended to the 1945 edition of his *The Meaning of Relativity*. In the paper he wrote in 1931, though, he not only conceded on the cosmological constant and the static model, but he also suggested exploring an alternative, one of Friedmann's models. He could have chosen Friedmann's ever-expanding universe, but he didn't. Instead, he fixed his attention on Friedmann's model of a closed, periodic (oscillatory) universe.

Einstein does not explain in his 1931 paper why he made this choice. He does not even mention the terms "periodic" or "oscillatory." However, his correspondence with other cosmologists preceding the publication can be found in the Einstein Archives, currently stored at Caltech. There one can find letters exchanged between Einstein and Richard Tolman, a Caltech chemist and cosmologist, indicating that Einstein's choice of a periodic universe was quite conscious. Einstein explains to Tolman that he is exploring an "oscillatory model" of the universe, in which the universe expands and contracts at regular intervals. The letters do not explain his motivation, but one can form a reasonable guess. Einstein had seen two important advantages of his static model. First, space and time exist forever, so the model is complete. There is no need to explain how the universe originates. Second, the universe is filled forever with matter and radiation. Einstein was disturbed by the idea that the unchecked expansion of the universe could cause the matter and radiation to become diluted. He was a devotee of the seventeenth-century philosopher Baruch Spinoza, who had declared that God would not make an empty world. Yet that is just what a monotonic universe (whether open or flat) would become.

Given Einstein's philosophical predilections, a periodic universe probably seemed the best alternative to his earlier static model. Space and time exist forever, and space always contains significant amounts of matter and radiation. Yet Einstein, in the same paper in which he raises the periodic possibility, unveiled a serious flaw. Using Hubble's measurement of the current expansion rate of the universe, he computed the age of the universe according to the periodic universe picture and discovered that it is less than the age of the Earth. Two years later, in Einstein's next paper on cosmology, written with Willem de Sitter, the periodic universe is not even mentioned. Instead, they consider a universe that is flat and expands forever.

As it turns out, about twenty years later, Caltech astronomer

Walter Baade showed that Hubble's value for the expansion rate of the universe was wrong by a factor of ten or more. To judge how fast the distant galaxies are receding from us, Hubble had relied on earlier measurements of Cepheid variable stars (described in chapter 2) that determined the relation between the rate at which the stars pulsed and their brightness. In 1952, Baade repeated the measurements using the powerful telescope at the Palomar Observatory and found a significant discrepancy, which increased the age of the universe by a factor of ten. Subsequent corrections led to further increases in the estimated age. If Einstein had known of these improved measurements of the expansion rate in 1931, his calculation of the age of the universe would have been at least twice the age of the Earth. It is interesting to imagine what Einstein would have written about the periodic universe in this circumstance.

Over the decades following Einstein's death, in 1955, cosmologists have obtained not only more accurate estimates of the expansion rate of the universe but also accurate estimates of the ages of stars. The result is that a new "age problem" emerged for the kind of periodic universe Friedmann and Einstein had in mind. The revised age of the universe, while significantly greater than the age of the Earth, is younger than the ages of the oldest stars. The introduction of dark energy, as occurs in the cyclic model, readjusts the estimated age of the universe according to the Friedmann equation and removes this conflict.

George Gamow, the Ukrainian-born theorist who, more than anyone else, was responsible for combining the Friedmann-Lemaître model with the laws of atomic and nuclear physics to create the big bang model recognized today, was also sensitive to the issue of whether the big bang was truly the beginning of the universe. In 1952, he published a beautifully written popular book called *The Creation of the Universe,* which recounted the history of the universe. The title of the book suggests that Gamow favored the idea of a universe created from nothing, but the story between the covers makes clear that he thinks otherwise. In fact, in his preface to the second printing in 1957, he apologizes for the title with his characteristic sense of humor:

> NOTE FOR THE SECOND PRINTING: In view of the objections raised by some reviewers concerning the use of the word "creation," it should be explained that the author understands this term, not in the sense of "making something out of nothing," but rather as "making something shapely out of shapelessness," as, for example, in the phrase "the latest creation of Parisian fashion."

Of course, the issue of a beginning was central in the minds of Gamow's chief critics and competitors, Fred Hoyle, Hermann Bondi, and Thomas Gold, who proposed the steady-state model of the universe in the 1940s. Hoyle, in particular, found the big bang abhorrent because he was vehemently antireligious and he thought the cosmological picture was disturbingly close to the biblical account. To avoid the bang, he and his collaborators were willing to contemplate the idea that matter and radiation are continually created throughout the universe in just such a way as to keep the density and temperature constant as the universe expands. This steady-state picture was the last stand for advocates of the unchanging universe concept, setting off a three-decade battle with proponents of the big bang model.

The debate was not confined to scientific circles. In 1950, long before the issue had been settled scientifically, Pope Pius XII weighed in with a Vatican encyclical that surely must have irritated Hoyle and disappointed Lemaître: "It would seem that present-day science, with one sweep back across the centuries, has succeeded in bearing witness to the august instant of the primordial Fiat Lux [Let there be light], when along with matter, there burst forth from nothing a sea of light and radiation, and the elements split and churned and formed into millions of galaxies."

Four Fatal Blows

As the battle between the steady-state and big bang models raged between leading cosmologists, the oscillatory model was largely ignored. The age problem that had deflected Einstein and Lemaître was not so much the issue. Instead, a series of more vexing problems seemed to totally smash any hope for a periodic universe.

The first new blow to the oscillatory model was dealt by Richard Tolman, the professor at Caltech in whom Einstein confided. Tolman was trained as a physical chemist but switched his focus to cosmology soon after taking his faculty position at Caltech in 1922. Tolman recognized that, according to the Friedmann-Lemaître picture, the universe can be treated like a giant chemistry experiment. It is filled with a uniform gas that cools and changes properties as it expands. The same laws that apply to conventional gases studied in the laboratory should apply to the cosmos, he reasoned. In 1934, he wrote a famous treatise, *Relativity, Thermodynamics and Cosmology,* which includes a description of many different types of cosmological models, including oscillatory ones. Although he had been sympathetic to the idea for many years (which is why he was corresponding with Einstein in 1930), his treatise presented a compelling argument against an oscillatory picture.

Tolman's argument rested on the second law of thermodynamics. Thermodynamics is the study of heat and how it flows from one body to another. At the heart of the subject lie two fundamental laws that are among the best known and respected in all of physics. The first law of thermodynamics says that energy is conserved, including any energy in the form of heat. Sum up the amount of energy that exists in all forms, and the sum does not change with time. The second law says that entropy always increases. Entropy is a measure of the disorder in a system, which amounts to counting the number of ways one can rearrange the constituents without changing the physical properties. Consider, for example, what happens to water molecules when the temperature is changed. In a frozen state, the molecules assemble into a crystalline array, a highly ordered arrangement with low entropy. If the molecules are heated, they dissociate into steam, in which the molecules move rapidly in random directions, a state with high entropy.

The second law says that in any system isolated from outside influence, entropy always increases. It may decrease in one part of the system, but then it must increase by even more somewhere else. For example, a freezer turns water into ice, reducing its entropy. But the generator that runs the freezer heats the air around it and creates more than enough entropy to compensate for the loss.

Another way of expressing the second law is that no machine is perfectly efficient because it must release heat and energy in a disordered form that cannot be 100 percent recycled. This rules out perpetual motion and is a serious threat to any kind of cyclic model of the universe. In order for a universe to have regularly repeating periods of evolution, it must be possible to restore its physical condition back to the way it was at an earlier time. But the second law seems to say that this is impossible, because the total entropy must increase from cycle to cycle.

In the oscillatory model considered by Friedmann, Einstein, and Tolman, most of the entropy is in the form of radiation. The amount of radiation increases as stars and galaxies form during the expanding phase of the model. The new radiation is then compressed, along with any radiation from previous cycles, into a tiny volume of space at the big crunch. As a result, the subsequent big bang begins with more radiation, and a greater concentration of entropy than the previous one. Using Friedmann's equation, Tolman showed that the higher temperature means the expansion rate after a bang is greater than it was the cycle before. If the expansion starts off faster, it takes longer before it halts and longer before the next big crunch. So the duration of each cycle is longer than the one before it.

Extrapolating backward in time, the cycles become shorter and shorter. The cycle duration shrinks so rapidly, in fact, that it reaches zero within a finite time. If the purpose of introducing the cyclic model is to avoid having a "beginning," Tolman's entropy argument showed that the oscillatory model fails. For the next seventy years, this argument dissuaded most cosmologists from pursuing the cyclic idea.

There remained a few ardent believers, most notably Robert Dicke from Princeton University, who led David Wilkinson, James

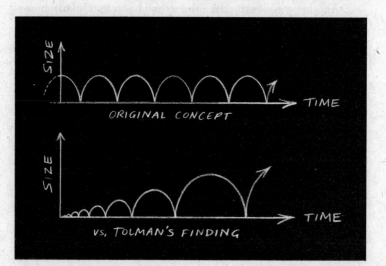

Peebles, and Peter Roll in interpreting Arno Penzias and Robert Wilson's discovery of cosmic background radiation in 1965. A curious feature of the classic paper by Dicke and his collaborators is that they explained the result not in terms of the conventional big bang model but, instead, in terms of an oscillatory model. This is because Dicke, like Lemaître and Einstein before him, did not think the idea of a "beginning" of the universe is plausible. The Russian cosmologist Yakov Zel'dovich wrote to Dicke afterward to chastise him for invoking the oscillatory model, repeating Tolman's argument based on the second law of thermodynamics. Dicke never amended the paper.

The discovery of the cosmic background radiation became widely interpreted in the minds of cosmologists, physicists, and the general public as the final proof that the universe had a definite "beginning." This is how it is described in nearly every popular account today, from elementary-school textbooks to graduate-school courses. Exactly how this interpretation became fixed, given the doubts expressed by Einstein, Friedmann, Lemaître, and Gamow, is not clear. Tolman's argument is one reason, but there may have been another significant psychological factor.

The discovery of the cosmic background radiation was the decisive observation in a decades-long battle between the steady-state and big bang models. One of the champions of the steady-state model, Fred Hoyle, was not only a world-renowned astronomer but also a highly effective public speaker and writer who did a lot to popularize astronomy and cosmology. To draw the strongest possible distinction between his own steady-state model and that of Friedmann, Lemaître, and Gamow, Hoyle liked to paint their model as requiring a "beginning," even though he must have known that the point was not settled. Hoyle even gave the concept a derisive name: "big bang." He was such a persuasive and well-known public figure that his description and terminology stuck. (Hoyle's needling may have been what spurred Gamow to write the note in the second edition of his popular book.) As a result, many years later, when the discovery of the cosmic background radiation upended the steady-state model, Hoyle's caricature had become the standard depiction.

A second blow to the oscillatory universe came in the 1950s and '60s, from theorists studying what happens to the universe as it contracts and approaches a big crunch. Russian physicists Vladimir Belinskii, Isaak Khalatnikov, and Evgenii Lifshitz showed that tiny differences in the rate of contraction become rapidly amplified, sending the universe into wild gyrations. Instead of contracting equally in all directions, space contracts along two randomly chosen directions and expands along the third, approaching a cigar shape. After a point, though, the situation suddenly switches, space contracting and expanding along different directions. These switches repeat over and over again, squeezing and stretching space much like dough kneaded in a large mixing machine. The squeezing and stretching differs from location to location, so the universe becomes highly inhomogeneous as the crunch approaches. The result, which Charles Misner of the University of Maryland dubbed the "chaotic mixmaster" universe, would cause space to emerge from the crunch with a highly nonuniform distribution of energy, completely inconsistent with the smooth universe observed today.

The problem could be avoided if the universe bounced from contraction to expansion before it got too small and the chaotic gyrations began. But by the 1960s and '70s, Stephen Hawking and Roger Penrose had shown that this is impossible if Einstein's theory of gravity is right. A shrinking universe, they showed, continues to shrink until space collapses to a state of infinite density known as the cosmic singularity. Similarly, 14 billion years ago, the universe must have emerged from such a cosmic singularity.

Hawking and Penrose's finding was widely interpreted as theoretical proof that space and time must have a "beginning." That interpretation, however, was never justified. What they really proved is that Einstein's equations become mathematically inconsistent at the big bang itself. This should be interpreted as a clarion call. Physicists must face up to improving Einstein's theory, if they want to describe the big bang. The improved theory might still predict a beginning, or it might allow a bounce, from contraction to expansion. If it is the latter, though, a method must be found to squelch the chaotic mixmaster behavior so that the universe emerges from a bounce in a state where space is uniform and flat.

In the 1990s, the old oscillatory models were in the news again. If they were dead before, based on theoretical reasoning, they were now killed twice more by new astronomical evidence.

First, a requirement of the old oscillatory models is that the expansion rate slows down as the universe expands, due to the gravitational self-attraction of the matter. Yet, as explained in chapter 2, there is now overwhelming evidence that the expansion of the universe is accelerating today.

Second, the oscillatory models require that the universe is closed and that the matter density is large enough for its gravitational attraction to cause the universe to collapse. By the mid-1990s, a host of new observations had shown unequivocally that the concentration of matter is far too low for this purpose. Also, space in a closed universe is curved, whereas the WMAP satellite experiment had shown that space is flat.

Thus, as the twenty-first century began, the oscillatory universe model, the predecessor of the new cyclic picture, appeared to have been killed four times over: once by the second law of thermodynamics; once due to chaotic mixmaster behavior; once by the observation of cosmic acceleration; and once because the universe has an insufficient amount of matter. There seemed to be little prospect of resuscitating an idea so fatally flawed.

Four Innovations

Four ideas have emerged since 1995 that intertwine to fend off all four fatal blows and to revive the concept of a regular repeating universe: extra dimensions, branes, dark energy, and dark energy decay. All four are incorporated in an essential way in the new cyclic model.

The extra dimensions arising from M theory enable one to build a new kind of cyclic picture. In the old approach, the usual three dimensions expand and contract as if the universe were breathing in and breathing out. During the contraction phase, space and everything contained within it is compressed to nearly zero size. In the new cyclic model, the usual three dimensions expand from one big bang to the next. The breathing in and out is done by the extra dimension. The combined effect is like pulling taffy. Three dimensions are stretched for a trillion years. Then the stretching rate slows, and the extra dimension is squeezed to nearly zero size and back. Then the three dimensions are stretched again.

Branes, key elements of M theory, are important for the new cyclic model because they determine where the matter and radiation are located within the many dimensions. In particular, matter and radiation exist only on the branes and not in between them. During the contraction phase, the space between the branes shrinks, but the branes themselves are always stretched and the matter and radiation along the branes is never compressed. If the universe becomes cold and matter is highly diluted (as it is today) before the contraction phase begins, it remains cold and diluted all the way to the big crunch because the branes remain expanded. Only the space between them shrinks to zero. As a result, the energy does not become infinitely concentrated at the big crunch the way it does in the earlier oscillatory models.

Dark energy, whose existence has been observationally established but which has no specific identity or purpose in the current inflationary model, plays three essential roles in the new cyclic model. First, as noted before, dark energy speeds up the expansion rate, causing the stars, galaxies, matter, and radiation produced since the last big crunch to spread out so thinly that the branes become nearly vacuous. A vacuum has a perfectly smooth distribution of energy. The accelerated stretching of the brane also flattens out any wrinkles that may have developed since the last collision and makes the branes parallel. This helps ensure that each new cycle begins with the same simple physical conditions as the cycle before.

The second role of dark energy is as a stabilizer. Suppose, for example, that some quantum or thermal fluctuation causes the branes to collide harder than usual. When the branes bounce apart, they might fly apart to a greater separation than usual. Now that the cycles are thrown off a little, you might worry that the next bounce will be even farther off. After a few collisions, the cyclic behavior could turn into random evolution.

Dark energy saves the day by acting like a shock absorber similar to the one on an automatic door closer. A pneumatic absorber consists of a piston in a tube. One may open the door halfway, three-quarters, or all the way. One may open it rapidly or slowly. But as the door begins to close, the piston pushes on the gas in the tube and heats it up. Some of the kinetic energy of the door is thereby converted into heat. By the time the door reaches its closing position, it is traveling at the same speed no matter how far or how fast it was opened. In a similar way, dark energy acts as damper for the colliding branes by converting the kinetic energy of the branes into gravitational energy. Under the influence of dark energy, the branes themselves stretch at an accelerating rate, but their motion along the extra dimension is slowed until it reaches the proper speed for cycling, even if they started off at the wrong separation.

The third role of dark energy is to shut itself off so that the universe can move on to the next stage of the cycle. Identifying dark energy with the potential energy associated with the springlike force drawing the branes together can automatically accomplish that goal. As the force slowly draws the branes toward each other, the potential energy decreases, eventually changing from positive to negative. Now the branes stop stretching and the dark energy phase comes to an end. The next stage occurs when the branes speed up and rush toward each other and the extra dimension contracts.

After a time, the brane kinetic energy and its negative potential energy nearly cancel each other, so the total energy is tiny. Yet according to Einstein's theory of general relativity, the outward pressure exerted by this tiny energy source depends on the kinetic minus the potential energy, which is enormous. The huge pressure, one can show, prevents small ripples in the branes from growing out of control.

The Last Question

The combination of extra dimensions, branes, and dark energy makes it possible to block all four blows that felled the oscillatory models of the 1920s and '30s and suggest a new kind of cyclic model. Consider first the fatal idea that there is too little matter in the universe to stop it from expanding forever. This is no problem for the new cyclic model because the cycling is not caused by the gravitational attraction of matter but, instead, by the springlike force between the branes. Furthermore, in the new cyclic universe model, which has low matter density plus dark energy today, the time elapsed since the big bang is increased compared to a universe with matter alone, enough to resolve the age problem that had plagued Einstein's periodic model.

Next we come to the discovery of cosmic acceleration. Accelerated expansion is a disaster for the old oscillatory models because it thins out the matter so much that the gravitational self-attraction of the matter becomes insignificant. But in the new cyclic model, cosmic acceleration is an absolutely required element. As described previously, the acceleration helps make the universe flat and uniform and is needed for keeping the cycling stable and on track.

As for the blow from chaotic mixmaster gyrations, which were thought to be inevitable in virtually any kind of contracting universe: a closer examination of the calculation done by Belinskii, Khalatnikov, and Lifshitz reveals an interesting loophole. They considered only examples where the pressure is smaller than the energy density. In the cyclic model, though, the reverse occurs: as we have already noted, the demise of dark energy is followed by a phase with large pressure and small energy density. Along with two graduate students, Daniel Wesley and Joel Erickson, we discovered that, much to our surprise, high pressure completely squelches chaotic mixmaster behavior and, instead, makes the branes smooth and flat.

This is the opposite of what everyone expected. When the cyclic model was first proposed, neither of us had been aware of the chaotic mixmaster problem. When we later found out about it, our expectation was that complications would have to be added to the model to evade the chaos, or perhaps that the chaos problem would be a fatal flaw. Instead, the model did not need even one iota of change. It was only necessary to discover what was there all along, waiting to be noticed. This eerie experience is typical of our experience with the cyclic model to date. Whenever a problem has arisen, it's turned out that the model already contains the ingredients necessary to address it. Not a single new element has been added to the picture since it was first envisioned in Finland.

This leaves only the last blow to consider, that of the second law of thermodynamics—the same problem Isaac Asimov tackled in his famous short story "The Last Question," published in 1956. That short story, one of his best known, was his favorite. According to the story, the Last Question is first posed in 2061 by two computer attendants who are babysitting the world's most powerful supercomputer, Multivac. Using the computer, humanity has just figured out how to directly tap the energy from the Sun. The two attendants are celebrating the event. One of them remarks that humanity now has enough energy to last forever into the future. Not so, says his dour partner. The energy will endure for only about ten billion years, he estimates, after which the Sun will run out of its nuclear fuel. It's a long time, but not forever.

So the two attendants begin to discuss what will happen when all the stars burn out, trillions of years hence. The second attendant argues that this will be the end of life. The burning of stars creates entropy. But once everything has burned up, the entropy reaches its maximum value. Since entropy only increases, that is the end of the line for life and the universe. His partner is less sure. Maybe humanity will find some way to reduce the entropy, he says. The second attendant thinks that is impossible, but they agree to put the question to Multivac. "How can the net amount of entropy of the universe be massively decreased?" they ask. Multivac does not respond.

The next scene takes place one million years later. Multivac has been replaced by what is now the most powerful computer in the galaxy, the Galactic AC, which continues to seek the answer to the same question. When queried, the computer responds that there is insufficient data to answer it.

A billion years hence and then ten trillion years hence, the scene repeats with ever more advanced computers. They continue to report that the data is insufficient. Finally, after the stars have burned out and humanity has passed into oblivion, a great cosmic computer, the Cosmic AC, continues to compute, moving slowly and inexorably toward an answer to the Last Question.

If the cyclic model is correct, the solution will come ten trillion years sooner than Asimov anticipated, and using string theory instead of a computer the size of the universe. According to the cyclic picture, the universe is infused with a sufficient amount of new matter and radiation during each collision to enable the creation of new galaxies, stars, and life from the collision of branes along an extra dimension.

Why doesn't the energy run out? Conservation of energy appears to be violated because each collision converts some fraction of the kinetic energy of the branes into matter and radiation. If the only energy drawing the branes together comes from the springlike force connecting them, the spring should wind down. The maximum separation between branes should decrease steadily as the bounces continue. However, there is a second force at work: gravity. During every cycle, a finite amount of gravitational energy is automatically converted into brane kinetic energy. This compensates for the loss of energy into radiation and matter that occurs in every cycle. In spite of the matter and radiation created at the collision, there is enough excess kinetic energy left for the branes to bounce back to their original positions and for the cycling to continue.

The process can repeat endlessly because gravity is a bottomless pit. Most familiar forms of energy are positive, and so there is a lower bound (zero) below which they cannot fall. Gravitational potential energy is negative, and there is no known limit as to how low it can go. It can decrease by a finite amount with each bounce and continue that way forever. The gravitational potential energy is not something that can be measured directly; thus it is not possible to detect that the potential energy is less after one bounce than it was after the bounce before. The only quantities that can be measured are the matter density, temperature, and expansion rate, and these quantities exactly repeat from bounce to bounce. Hence, an observer interprets the universe as being exactly cyclic. Behind the scenes, though, gravity is acting like an engine that keeps supplying more energy to keep the cycles going while respecting the conservation of energy.

As for avoiding Tolman's entropy problem, the solution depends on branes, extra dimensions, and gravity, combined with a more refined understanding of Tolman's argument. As Tolman maintained, the entropy must increase from cycle to cycle according to the second law of thermodynamics, and this remains true for the new cyclic model. But increasing the total entropy is *not* what led to Tolman's ever longer cycles. Closer examination of his argument reveals that the problem was increasing the entropy density. In the particular model that Tolman considered, all the dimensions of space contract as the universe approaches the crunch. As a result, the new entropy plus any old entropy become highly concentrated at the crunch, resulting in a higher entropy density than in the cycle before. According to Einstein's equations of general relativity, it is the greater entropy density that produces the bigger bounce. If the entropy had somehow remained spread out, each bounce could have been identical to the one before.

In the new cyclic model, only the extra dimension contracts. The entropy is created on the branes—for example, at the brane collision or when galaxies and stars form—and then spreads thinly during the rest of the expansion epoch. The accelerated expansion due to dark energy is especially effective in enabling gravity to make plenty of space to accommodate the new entropy and keep its concentration low. Then the branes continue to expand during the contraction phase—only the extra dimension contracts—so the entropy on the branes is never concentrated. When the branes collide and hot new matter and radiation are created, any preexisting entropy is exponentially diluted. Only the new matter and radiation are concentrated enough to affect the expansion. And since the amount is the same as in the cycle before, the duration of the next cycle is the same, as well.

In Asimov's imaginative story, the way to evade the second law of thermodynamics and bring the universe back to life was eventually discovered by advanced computers long after humanity had disappeared and the universe was nearly vacuous: "For another timeless interval, AC thought how best to do this. Carefully, AC organized the program. The consciousness of AC encompassed all of what had once been a Universe and brooded over what was now Chaos. Step by step, it must be done. And AC said, 'LET THERE BE LIGHT!' And there was light—"

The cyclic model is based on physicists' best efforts to describe and unify all the laws of nature. It has been shown, without the aid of a futuristic computer, that a combination of branes and an extra dimension, with regular assists from gravity and dark energy, can cause the universe to repeatedly replenish itself with galaxies, stars, and life at regular intervals while always obeying the second law of thermodynamics. As it turns out, the transition from a nearly empty universe to one filled with hot radiation is not all that far from the scenario Asimov envisioned. The collision between two branes would produce a searing white light, signaling the beginning of a new cycle of cosmic evolution.

chapter nine

Seeing Is Believing

The true method of knowledge is experiment.

—William Blake, "All Religions Are One"

Theoretical physicists across the globe may labor for years to develop the cyclic model more fully. The theory may successfully explain many of the detailed features of the universe and may provide a satisfying answer to the Last Question. Developments in string theory may provide added support. The theory may turn out to be mathematically beautiful and philosophically appealing. And then, after all that, the theorists may wake up one morning, connect to the Internet, and learn about a new observation that completely kills the idea. Mathematical beauty and philosophical appeal are useful guides, but they can be overcome by a single decisive experiment. Being a theoretical physicist means being willing to take this kind of high-stakes gamble and accept the consequences either way. You must enjoy the challenge and be prepared for the thrills and spills.

Thursday, March 16, 2006, was just such a live-or-die moment for the cyclic model. Neil had flown across the Atlantic the evening before, a trip planned several months earlier so that the two of us could get together for a week to work on the manuscript of this book. By pure coincidence, three days earlier the WMAP team had announced that on Thursday they would release new results from their first three years of observations. Unlike the first announcement in 2003, described in chapter 1, this time we were both on the same continent and in the same room, listening to the announcement. The WMAP satellite had maintained its orbit a million miles from the Earth, and the WMAP team had continued to gather cosmic background radiation from all directions in the sky. Through painstaking analysis, the WMAP team had used the added observations to significantly improve on their previous results and to reach a new milestone in testing cosmological models.

Two of the leading WMAP team members were Lyman Page and David Spergel, both at Princeton University. Page's office is just three doors down the hall from Paul's, in the Physics Department, and Spergel's is in the Department of Astrophysical Sciences, a hundred yards away. But the WMAP team had a strict code of silence. There was to be no release of information until March 16, and Page and Spergel had remained true to the code. They had graciously agreed to discuss their results before the entire Princeton scientific community on Thursday afternoon. A lecture hall that was twice the capacity of the one used for the first WMAP announcement was reserved, and Paul was asked to introduce the discussion.

Still, when Neil packed his bags in Cambridge, no one other than the team members had any idea what the new results indicated. The new WMAP measurements might annihilate the cyclic model, the inflationary model, or both. That would mean a short trip . . . and a very different book.

Testing Inflation

Although neither of us knew what the WMAP team would present, we knew what to be watching for. In the early 1980s, theoretical cosmologists studying the inflationary model had identified six key predictions. Testing any one of these is a technological challenge, so anytime a prediction is verified, it is a milestone achievement in establishing the inflationary hypothesis. Some of these tests have been described in previous chapters. The first milestone was achieved in 1992 by the Cosmic Background Explorer (COBE) satellite, which gave the first indication that the cosmic background radiation temperature has a *nearly* scale-invariant variation across the sky, as pictured on page 57. A second milestone was showing that space is flat rather than curved. Between 1997 and 2003, numerous experiments, ranging from Page's ground-based detector on Cerro Toco, in Chile, to the WMAP satellite, verified this prediction with progressively improving degrees of accuracy. The third milestone was confirming the inflationary prediction that ordinary matter and dark matter in the early universe were distributed in the same up-and-down pattern as the cosmic background radiation temperature, a condition known as *adiabaticity*. And the fourth was showing that this distribution has the random noise characteristics that inflation predicts, a condition known as a *gaussianity*. The WMAP satellite reached these milestones in its first year.

Two milestones remained. After two additional years of satellite measurements, the WMAP team could be in a position to say something about one or both. The fifth milestone was detecting a "tilt" in the amplitudes of energy density variations. Tilt refers to a systematic deviation from perfect scale-invariance. Perfect scale-invariance, as explained in chapter 3, means that the density of matter and radiation can be expressed as a sum of waves with similar heights, independent of the wavelength. Inflation predicts that the heights should get slightly smaller as the wavelength decreases. Cosmologists call this red tilt, as distinguished from blue tilt, in which the heights increase systematically as wavelength decreases. The red tilt occurs because the waves with smaller wavelengths are produced closer to the end of inflation when the inflationary energy is decreasing and creating less quantum jitter. The effect is so small that the WMAP team had not been able to confirm or deny it based on the first years of observations, but it was clear they had a chance after two more years of observations. With this thought in mind, Paul's graduate student Latham Boyle had worked with Neil and Paul to carefully examine just how much tilt the inflationary picture naturally predicts. Boyle's analysis showed that a red tilt of a few percent is expected.

In a series of parallel papers with Justin Khoury, the same student of Paul's who had helped develop the ekpyrotic model, and Boyle, we had also checked the predictions for the cyclic model for the first five milestones and showed that, remarkably, they are nearly identical to those for inflation. Despite the basic differences between the inflationary and the cyclic pictures, there are surprising mathematical symmetries that ensure that the two agree very closely (at the level measurable by WMAP) on the *near scale-invariance*, the *flatness*, the *adiabaticity* and *gaussianity*, and even on the *tilt*.

This meant that the WMAP team, after three more years of analysis, controlled the fate of both models. One possibility was that both models could be eliminated in one stroke—for example, if the observations conflicted with one or more of the five predictions above. If the models failed, surely some theorists would rush to add complications to make them fit the latest data. But these fixes would make the theories much less attractive, and could well spell the beginning of the end. If the models passed, there was a sixth milestone test to consider. The two models give vastly different predictions for the production of cosmic gravitational waves. The detection of these gravitational waves, therefore, could determine which model was correct.

Five out of Six Is Not Enough

Although the notice had come at the last minute and some students and faculty had left for spring break, nearly all 350 seats in McDonnell Lecture Hall were occupied when Page rose to speak. He began with an explanation of how the experiment worked, beginning from basic principles and walking the audience through the instrumentation and observations. His presentation included a series of improved WMAP images of the temperature variation across the early universe. The awesome display demonstrated clearly that every instrument on board had been thoroughly analyzed and was working to perfection. He also presented a new kind of image called a polarization map, which will be described later in this chapter. At the halfway point, Spergel explained what the new results meant for cosmology. He showed how this data strengthened the case for the first four milestones and improved the quantitative measurements of the ordinary matter, dark matter, and dark energy densities. And, then, sure enough, Spergel presented evidence for the fifth milestone: WMAP now had evidence for red tilt, he explained, and with a value right down the middle of the expectation for both inflation and cyclic models. Recognizing that they were witnessing history in the making, everyone in the auditorium broke into thunderous, long-lasting applause.

The discovery of red tilt had pushed cosmology into the realm of testing the role of fundamental physics in detail. Reaching the first four milestones was impressive and important, but, historically, cosmologists had guessed that these conditions are likely on the basis of simple astronomical observations and heuristic reasoning, without invoking particle physics or string theory. For example, as mentioned in chapter 3, Harrison, Peebles, and Zel'dovich had anticipated the fact that the primordial density variations are scale-invariant. They came to this idea by working backward from the fact that today's universe is reasonably smooth on large scales, and is not composed almost entirely of black holes on small scales. This gave a constraint on what kinds of fluctuations must have existed initially. They concluded that scale-invariance is the most plausible scenario, although they had no physical mechanism to explain why it should be that way. A similar story applies to milestones two to four. For this reason, some astronomers had discounted ideas like inflation or cycling, figuring that they had not predicted anything that had not been inferred already through classical physics and logical reasoning.

But no one had anticipated tilt. Tilt could be created by the quantum fluctuations of fields, strings, or branes in an expanding universe where the rate of expansion is slowing down, or in an ekpyrotic contracting universe where the contraction of the extra dimension is speeding up as the branes get closer. It is an effect that arises naturally from theories that combine fundamental physics and cosmology. The effect is too small to have been guessed from working backward from the general properties of today's universe. With the detection of tilt, the linkage between cosmology and fundamental physics is forged.

The press release by NASA that accompanied the WMAP three-year announcement spoke only of new support for the inflationary model, and that is the story that most newspapers carried. This was understandable since the cyclic picture was newer and much less known, especially to astronomers. But there was a story behind the story.

By confirming the predictions of the *simplest* inflationary models, the WMAP three-year results greatly increased the pressure to pursue the sixth milestone, reversing the trend over the years preceding the announcement. Prior to the WMAP announcement, many theorists had suggested that passing only four or maybe five out of six milestones might be enough to conclude that inflation is correct. They argued that the tests were not so important because it is possible to design inflationary models that do not satisfy them. To make their case, they constructed complicated models for inflationary energy with extra components and fine-tunings that would match the first five milestones, say, but evade the sixth milestone by producing gravitational waves too weak to be detected with foreseeable technology.

Designing an inflationary model to avoid a milestone test is worrisome, though, because the same approach can be used to skirt any of the other milestone tests, or any combination of them. Isn't it questionable practice to declare that inflation is confirmed if it passes one milestone test, and still valid if it fails another? By showing that the simplest inflationary models pass the first five milestones, WMAP made it very awkward, if not illogical, to introduce fine-tuning to evade the sixth. The same reasoning applies for the cyclic model. In this way, the new WMAP results certified the simplest inflationary and cyclic models and cleared the field for a head-to-head competition between the two, to be settled by the sixth milestone test.

Making Waves

The ideal way of distinguishing the inflationary and cyclic pictures would be to look back in time to see what actually occurred a few instants after the big bang. While astronomers can look sequentially further back in time by looking deeper into space, the problem is that the early universe is filled with dense plasma that blocks any light emitted before the 380,000-year mark from ever reaching the Earth. To see through the barrier, cosmologists need to use a far more ethereal form of radiation, one able to pass unhindered through the dense early universe and reach their detectors. Fortunately, just such a ghostly source of radiation exists, one of the weakest, most ancient, and evanescent entities in the universe: cosmic gravitational waves. Gravitational waves are the key to the sixth milestone test, which distinguishes the inflationary model from the cyclic model.

Gravitational waves are distortions of space that travel through the universe like ripples on the surface of a pond. As the waves move through space at the speed of light, they cause space to alternate back and forth between squeezing along one direction and stretching along a perpendicular direction, where both of these directions are at right angles to the motion of the wave. To picture this effect, consider what would happen if a gravitational wave were to pass through Radio City Music Hall during a performance of the Rockettes, the troupe famous for its long row of dancers with identical heights moving in perfect synchrony.

We can represent the wave itself as a Slinky that is squeezed and stretched according to the way a gravitational wave distorts space. At some points along the wave, space (and the Slinky) stretch up and down and squeeze front and back. Then, farther down the line, space (and the Slinky) squeeze up and down and stretch front to back, and so on. As the gravitational wave passes through the line of Rockettes, then, it first increases the height of a dancer and squeezes her front to back, making her appear taller and thinner than usual. Then, just as she is becoming pleased by that outcome, the next part of the wave comes along and makes her short and fat. The pattern repeats at regular intervals as the wave continues along. Although the net effect is

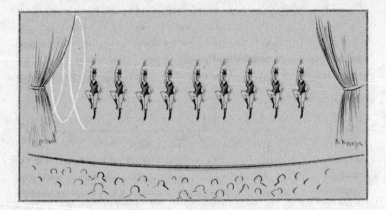

nil, the wave would surely cause consternation and chaos along the usually disciplined line of dancers if the effect were as large as the cartoon suggests.

In actuality, the waves are so weak that their effects are impossible to detect without a highly sensitive instrument. It's a good thing, too, since all of us are constantly being bombarded by gravitational waves. Any matter or energy that accelerates, wiggles, sloshes back and forth, or makes circles creates gravitational waves. Pick up this book and move it to and fro, and you have created gravitational waves that travel outward in all directions at the speed of light. At any given time, you are being hit by gravitational waves produced throughout the universe, including waves produced nearly 14 billion years ago, around the time of the big bang. If you could measure and decipher the signal, it would provide a chronicle of the history of the universe. However, the total amount of squeezing and stretching as the waves reach the Earth is typically much less than the width of an atomic nucleus. Only a highly sensitive instrument can detect this tiny signal.

In the inflationary model, gravitational waves are generated through quantum jitters of microscopic regions of space. Random quantum jitters create tiny warps and ripples in space all the time—even now, not just during inflation. But normally they come and go so quickly that they leave no long-term vestiges. The warps created during inflation are different because they are rapidly stretched to extraordinary sizes and become long-lasting distortions of space. By the end of inflation, all the warps and wiggles, ranging from those that have been stretched a lot to those that are still microscopic in size, have roughly the same height and depth, and all contribute to the distortion of space.

The situation at the end of inflation is similar to stretching out an elastic sheet so that it is very smooth and flat and then having a group of friends grab hold of one part or another in the middle to create little hills and valleys. Next, imagine having them let go. Where the elastic has been plucked, it snaps back and begins to oscillate up and down, creating ripples that travel across the surface.

Likewise, at the end of inflation, the warped regions begin to undulate up and down and generate gravitational waves that travel in all directions. Small-wavelength gravitational waves are set in motion first, and then progressively longer-wavelength gravitational waves start moving as the universe evolves. Today, the wavelengths range from a few meters to billions of light-years. The waves, once they start to travel through the universe, begin with nearly the same height for all wavelengths, a result reminiscent of the scale-invariant variations in energy density depicted on page 57. Consequently, the sixth milestone test for inflation is to search for a scale-invariant spectrum of gravitational waves.

The story for the cyclic model is completely different, for two reasons. First, the energy density of the universe during the phase when long-wavelength gravitational waves are generated is minuscule for the cyclic universe compared to the inflationary case. With less energy density, there is less gravitation to convert quantum jitters into waves. Second, the gravitational waves produced in the cyclic universe are not scale-invariant. Instead, their amplitude increases sharply as their wavelength decreases. This is because the gravitational waves are generated as the branes accelerate toward one another before the big crunch. The acceleration causes the quantum jitters of space to increase as the branes approach, which enhances the height of the smaller-wavelength gravitational waves generated during the last instants before collision. The result is a cyclic spectrum of gravitational waves that cannot be confused with the inflationary prediction.

The Sixth Milestone

The sixth milestone test is to search for cosmic gravitational waves and determine if they agree with the inflationary prediction. If the test is passed, this will be compelling evidence that the gravitational waves, the temperature fluctuations seen by WMAP, and the initial uniformities that seeded galaxy formation were all created in a period of high-energy, ultrarapid inflation. The discovery would also crush any hope for the cyclic model. Conversely, failing to find the gravitational waves would suggest that their heights are too small to measure and that the WMAP temperature fluctuations and the seeds for galaxies were created in a gentler process, like the one proposed in the cyclic model.

The gravitational waves from inflation are very feeble, and so reaching the sixth milestone is challenging. Nevertheless, as Page and Spergel explained in their presentation, the WMAP team already had something to report about this issue, and cosmologists worldwide have aimed their research at performing this decisive test.

WMAP is relevant because gravitational waves leave a distinctive imprint on the cosmic background radiation pattern. First, gravitational waves traveling across the universe distort the distance to the plasma that emitted the cosmic background radiation. The effect is to brighten or dim the radiation according to whether the path along which it travels is shrunk or stretched by the gravitational waves. The brightening and dimming creates hot and cold spots in the WMAP image on top of those caused by the variations in energy density.

The effects of gravitational waves on the WMAP image can be distinguished from the effects of energy density variations through a statistical analysis of the pattern of hot and cold spots. For example, one approach is to count how the number of spots varies with size, where the size is measured by the number of degrees a spot subtends on the sky. The gravitational waves generated during inflation should produce a pattern with a nearly equal number of detectable spots of each size for sizes ranging from two degrees and upward. Variations in energy density produce a different pattern of hot spots and cold spots spanning all angles. By comparing the number of spots of various sizes, cosmologists can disentangle the gravitational wave and energy density contributions. The simplest inflationary models predict that gravitational waves should be responsible for somewhere between 10 and 40 percent of the hot and cold spots spanning more than two degrees. Using this and other more sophisticated statistical tests, the WMAP team found no gravitational signal down to the level of 35 percent, ruling out some simple inflationary models but leaving many others.

A more sensitive test relies on the effects of gravitational waves on the *polarization* of the cosmic background radiation pattern. As explained in chapter 4, light is an electromagnetic wave whose electric and magnetic fields oscillate in two perpendicular directions, both at right angles to the direction of the wave. The polarization of the wave is defined as the line along which the electric field oscillates. In the figure on page 207, for example, only the electric field is shown. The incoming wave in the top panel of the drawing represents light in which the electric field oscillates up and down as the light wave travels along a horizontal axis. So the incoming light is said to be "up-down polarized." Similarly, the middle panel shows a wave that is left-right polarized, and the bottom panel shows light that is unpolarized, meaning that it contains a mix of different polarizations.

If light with up-down polarization scatters from an electron in its path, as shown in the top panel, part of the wave travels straight ahead with up-down polarization, and part of the wave scatters to the right at ninety degrees, keeping the same up-down polarization. (The wave could scatter both right and left, but only a scattering to the right is shown to keep the picture simple.) Note, though, that no light is scattered by ninety degrees up or down. This is because the polarization must be perpendicular to the direction of motion, so that up-down polarized light cannot travel in the up-down direction. Similarly, the middle panel shows that the left-right polarized light can scatter up and down, but not left or right. Finally, consider a mixture of waves with up-down, left-right, and other polarizations, as shown in the bottom panel. The part of the wave that travels

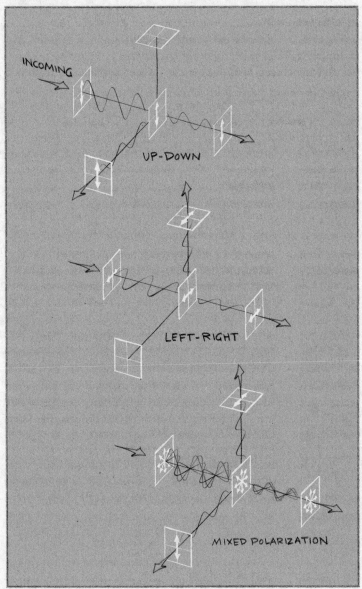

straight ahead remains a mixture. But the light that scatters by ninety degrees is polarized. In other words, scattering can turn unpolarized light to polarized light.

The polarization can be seen by looking at the wave itself or at the double-headed arrows drawn on the white planes. The double-headed arrows are a useful shorthand for indicating the polarization direction coming in or out. To further simplify the picture, one can drop the arrow heads and use only a solid bar to indicate the polarization (as in the next figure).

Light from many sources, including the Sun, an incandescent bulb, and the hot plasma in the early universe, is unpolarized. When the unpolarized light scatters off matter, though, the outgoing radiation in some directions is polarized, as shown in the figure on page 207. For example, light from the Sun scattering by roughly ninety degrees from molecules in the atmosphere arrives at the eye highly polarized (though not perfectly so). That is why sunglasses often contain polarized lenses designed to block the predominant polarization and to let through only the small amount of light with the opposite polarization.

In a similar way, the cosmic background radiation became polarized when it scattered for the last time off the hot gas that filled the universe 14 billion years ago. If the universe were perfectly uniform, there would be no net polarization because the polarizations produced by light rays hitting the hot gas from different directions would cancel one another out. Because the hot plasma has some nonuniformity, the cancellation is imperfect and the scattered light is polarized. By the time it reaches us, the wavelength of the radiation has been stretched by the expansion of the universe, transforming the red-hot light into invisible microwaves, but the polarization is unchanged. The polarization is a useful diagnostic for testing cosmological models.

The tiny nonuniformities causing polarization come from the two different effects that have already been described: energy density fluctuations and gravitational waves. The energy density fluctuations create a roughly spherical pattern of inflow of plasma leading to a characteristic pattern of polarization known as "E-mode" (because of its resemblance to the electric field pattern produced by a set of charged particles). John Carlstrom from the University of Chicago and his team of collaborators first detected this kind of polarization in 2000 using DASI (Degree Angular Scale Interferometer), an array of thirteen detectors constructed at the South Pole.

Gravitational waves are an entirely different source of temperature variation and polarization. They create hot spots by squeezing space in one direction and stretching it in a perpendicular direction, creating a strongly aspherical pattern of flow in the plasma. The corresponding polarization pattern is complex, involving not only E-mode but also a twisting pattern called B-mode (resembling the patterns of magnetic force fields, which physicists traditionally signify with the letter B).

In his talk, Page proudly presented the first full-sky polarization

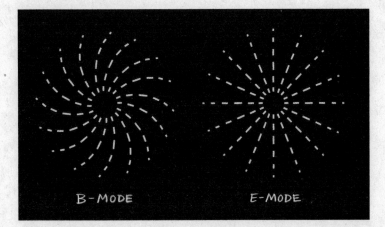

map ever made of the cosmic background radiation during his part of the talk. Superimposed on the WMAP image were lines showing the polarization directions, analogous to the bars in the previous figure. (They look like arcs in the figure only because the image is made by projecting the spherical sky onto a flat oval.)

Since energy density fluctuations produce only E-mode polarization but gravitational waves produce a mixture of both E- and B-mode polarizations, a question is raised: Is there any B-mode polarization in the WMAP pattern? If so, this verifies the presence of gravitational waves and inflation alone passes the sixth milestone test. It is impossible to tell by eye, but there is a well-known, precise mathematical method for separating the two. If there is even a tiny component of B-mode, the test can detect it, in principle, unless other sources of polarization from within our galaxy obscure the signal.

After displaying the map, Page hastened to report, "The pattern we have measured is pure E-mode, consistent with both the inflationary and ekpyrotic predictions." More specifically, the test showed that the gravitational wave contribution to the hot and cold spots had to be less than 28 percent, ruling out still more inflationary models.

Spergel followed up in his half of the talk with an analysis of

specific types of simple inflationary models. For the experts in the audience, the impression was unmistakable. Although the team was emphasizing the agreement with the inflationary picture, by failing to find evidence for gravitational waves, they had actually ruled out some of the most promising models. Now a modest improvement in B-mode measurements might be enough to discover gravitational waves, if they are there, or to push inflationary proponents into a corner where they would have to introduce further special fine-tunings.

The Race Is On

After the applause died down, there was time for questions from the audience. As the end of the question period approached, Page was asked, "You have presented the first map ever made that displays the polarization of the cosmic background radiation across the sky, demonstrating that it is possible to overcome the obscuring effects due to the Milky Way and intergalactic dust to some degree. So you know what we are up against. With this experience, how much further do you think we can push to search for B-modes and what is the best experimental approach?"

The audience leaned forward a bit to hear the answer, since Page was known to be a meticulous scientist whose pronouncements on experiments and their prospects was based on careful consideration and conservative judgment. Page's answer was crisp and unequivocal: "We can push the measurements from twenty-eight percent to just one percent using a new satellite dedicated to the purpose."

The statement was the rhetorical equivalent of firing a starting gun. It confirmed that the race to the sixth milestone was on. In trying to construct their polarization map, the WMAP team could have found insurmountable problems. After all, the polarization is caused by a scattering of light that occurred over 13 billion light-years away. For that light to reach us, it must travel an enormous distance and pass dust and other material that can change its polarization. Also, light from stars, hot dust, and molecular clouds adds to the cosmic background radiation and tends to drown out the signal being sought. These effects are real hazards that the WMAP scientists had been combating since their first-year announcement in 2003, and the team could have discovered that further improvement is impossible. Instead, Page's statement meant that WMAP had not found any roadblock to reaching the 1 percent level, a standard more than sufficient to check the predictions of the simplest inflationary models. Page's smile as he finished the sentence made clear that he was already contemplating the prospect.

To reach the 1 percent level, Page was calling for a satellite whose instruments and flight path are specifically designed to measure the polarization pattern across the entire sky without hindrance by the atmosphere, the Earth, or the Sun. An example would be a satellite sent to the same location as WMAP but equipped with an accurate polarimeter, a device that measures polarization. The instrument could produce a precise, high-resolution polarization map to replace the crude one that Page had shown in his talk, in the same way that WMAP had surpassed the coarse-grained temperature map produced by the COBE satellite experiment a decade earlier. Experimental cosmologists had already begun to speculate about plans for such a mission, tentatively dubbed CMBPOL (for Cosmic Microwave Background POLarization satellite) and also called the Inflation Probe. However, until the WMAP three-year results, they could not be confident that the mission was technically feasible. Now the situation had become clear. The critical sixth-milestone test distinguishing the inflationary and cyclic models can be done.

What is less clear even now is how long it will take to get approval for the project from a governmental space agency. In the meantime, experimentalists are not content to wait for the next space mission. If the inflationary picture is right, as many believe, then WMAP may have barely missed discovering cosmic gravitational waves. There is a chance that within the next few years a CMBPOL mission could be scooped by other experiments that make only a modest improvement in the B-mode search. There is a small chance that WMAP itself will make the breakthrough as it continues to gather data for the next three years or more, although it is near the limits of its capacity so far as the B-mode search is concerned.

In 2008, the European Space Agency (ESA) will launch the Planck satellite experiment; it will begin to report results a year or so afterward. This satellite is designed to improve measurements of the temperature variations of the cosmic background radiation to obtain a map with even higher resolution than the WMAP image. The satellite will also measure the polarization of the cosmic background radiation and could significantly improve on the WMAP limit. In 2001, shortly after the first paper on colliding branes appeared, we both spoke about the ekpyrotic model at a major conference in Cambridge called M Theory Cosmology. Stephen Hawking, the preeminent theorist who had pioneered studies of the initial singularity and made important contributions to inflationary theory, is a close colleague of Neil's and was in the audience. At the end of Neil's talk, Hawking made a public bet that the Planck satellite would detect the gravitational waves from inflation and rule out the ekpyrotic model. Neil readily accepted the wager, at even odds, for any amount Stephen would care to mention. Perhaps out of a gracious unwillingness to bankrupt Neil, Stephen has resisted naming financial terms. But he stands by the bet, and a suitable prize will be negotiated by the time the Planck satellite flies.

At the same time, many ground-based and balloon-based projects are in the planning or construction stages. Known by the acronyms SPIDER, SPUD, EBEX, PolarBear, QUIET, BICEP, and CLOVER, and involving teams of scientists from dozens of research institutes around the world, these experiments are racing to detect the polarization signal from gravitational waves produced in the early universe. They have the disadvantages, compared to the Planck satellite, of having to look through the earth's atmosphere and of seeing only a portion of the sky. On the other hand, they are employing instruments and designs specifically suited to detecting polarization, so they have a plausible chance of detecting a B-signal too weak even for the Planck satellite to sense.

The experiments are both technically and physically challenging. An example is the QUIET experiment being constructed by a team of physicists from about ten universities across the United States together with NASA's Goddard Space Flight Center and Jet Propulsion Laboratory. QUIET, which stands for Q/U Imaging ExperimenT (*Q/U* is physics code for "polarization"), will take advantage of a breakthrough in polarimeter design made at the Jet Propulsion Laboratory that makes highly sensitive polarization measurements possible. The detectors will be mounted on a series of telescopes in the Chajnantor scientific reserve at an altitude of 5,080 meters in the middle of Chile's high-altitude Atacama Desert. A photograph of the site reminds one of a lunar landscape. It was chosen for its altitude and lack of humidity, which will greatly reduce atmospheric interference. The project is limited to measuring the polarization over a few degrees of the sky, which may not be enough to detect an inflationary B-mode signal. However, advanced experiments like this one and those being conducted at the South Pole and in balloon flights around the coast of Antarctica, even if they fall short of detecting the signal, are all playing a critical role in developing and testing the new technologies needed for the ultimate polarization experiment: the space-based Inflation

Probe satellite, which is the best hope over the next decade or two for distinguishing between the inflationary and the cyclic models.

In addition to seeking the polarization imprint of gravitational waves on the cosmic background radiation, physicists will try to distinguish the inflationary and cyclic models by detecting the gravitational waves directly. For this purpose, instruments are needed that can sense the infinitesimal squeezing and stretching of space that occurs when a gravitational wave passes by. The required technology is not available today and will probably not be available for several decades. Nevertheless, direct detection could play an important role in corroborating and improving on the polarization measurements.

Even if the inflationary model is correct, a number of hurdles must be crossed before direct detection of the gravitational waves from inflation is conceivable. The first step is to find evidence for gravitational waves from stronger sources. This has already been achieved in the case of binary star systems in which two stars rapidly revolve about each other and one of the stars is a pulsar. Pulsars are rapidly spinning neutron stars that emit intense, narrow beams of electromagnetic radiation, which rotate like giant lighthouse beams as the pulsars spin. The pulsars that are detected are those whose beams strike the Earth at regular intervals of time, just like the ticks of a clock. Some pulsars are members of binary star systems in which they orbit another ordinary star. The detailed observations of the regular pulses can be used to determine what is happening to the pulsar's orbit. Careful studies by Joseph Taylor at Princeton University and his collaborators have shown that the orbits are decaying at the precise rate expected if the binary system is emitting gravitational waves in accordance with Einstein's general theory of relativity. The agreement is an indirect proof that gravitational waves really exist, an achievement recognized by the Nobel Prize awarded to Taylor and his former student Russell Hulse in 1993.

The second step is direct detection of gravitational waves from the binary star systems and colliding black holes. These waves should be much stronger than those from inflation. The Laser Interferometer Gravitational-Wave Observatory (LIGO) is currently being constructed for this purpose. On a flat plain in Hanford, Washington, and in the midst of a logging forest near Livingston, Louisiana, lie two detectors, each with two arms about four kilometers long that have been joined together in the shape of a giant L. Massive mirrors have been placed at the corner and the two ends of each L, and a highly intense laser beam is sent back and forth along both legs. The beams are precisely tuned so that each time the two laser beams meet at the corner, the light adds together destructively. That is, the crests of the light waves traveling along one leg coincide with the wave troughs from the other leg, causing the two signals to exactly cancel each other. Physicists call the effect *interference*. When a gravitational wave comes along, it squeezes one leg and stretches the other. Now the light-wave crests and troughs do not match precisely and the two signals do not precisely cancel. As a result, the combined light produces a detectable signal as the gravitational disturbance passes. If the wavelength of the light is very short, the crests are tightly spaced and even a tiny shift between crests and troughs can be detected.

The LIGO L's are designed to detect changes of 10^{-16} centimeters, or about one hundred millionth of the diameter of a hydrogen atom. Tiny variations of this magnitude can be caused by many ordinary phenomena, including microearthquakes, waves impinging on the shore, and the felling of nearby trees. By constructing two identical distant laboratories and searching for coincident signals, scientists ef-

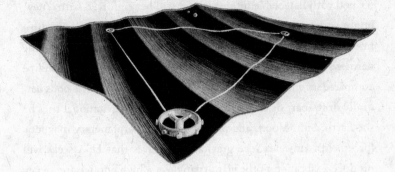

fectively screen out such local sources. What remains is a signal from gravitational waves whose kilometer-scale wavelengths are comparable to the lengths of the arms of the L. Over this range of wavelengths, LIGO is probably not sensitive enough to detect the weak gravitational signal predicted by inflationary models, although its designers are looking to improve the experiment so that this will be possible. Even if it falls short, LIGO, and its European cousin, VIRGO, will be important for establishing the field of gravitational wave detection and developing technologies that will help subsequent detectors search for the gravitational waves predicted by inflationary models.

After LIGO, the next gravitational wave observatory will be based in space. The Laser Interferometer Space Antenna (LISA) will consist of three free-flying spacecraft arranged in an equilateral triangle about three million miles on a side. Each spacecraft will be a pod with sensors and thrusters enabling its position relative to the other pods to be tracked and maintained with ultrahigh precision. In the heart of each pod will be a freely floating, shiny metal cube about four centimeters on each side. The faces of the cube will act as mirrors to reflect laser beams fired from the master pod and bounced off the cubes inside the other two pods. When a beam arrives at one of the distant pods, it will be too weak to make the trip back to the master pod with detectable intensity, so it will be reinforced with a new, bright laser beam whose crests and troughs will be carefully synchronized with the crests and troughs of the original beam. When the new beams complete the trip back to the master pod, they will be compared, as with LIGO, to see if the pods have shifted. The pods naturally drift apart or closer together due to the gravitational tugs by the Earth, Sun, Moon, and planets and by interplanetary magnetic force fields. However, the gravitational waves that LISA seeks will produce a vibration of the giant triangle at a high enough frequency that it will be clearly distinguishable from these more mundane sources of motion.

The LISA mission is designed to search for gravitational waves from distant colliding neutron stars or black holes that are either too far away or not violent enough for LIGO to detect. It should also see bursts of waves from the collision of super-massive black holes, over a million times more massive than the Sun, in distant galaxies. But it would take an instrument with over a thousand times LISA's sensitivity to see cosmic gravitational waves.

Always thinking ahead, physicists have already begun to plan for the ultimate step: the direct detection of gravitational waves predicted by the inflationary model or compelling evidence that they do not exist. This satellite concept, dubbed the Big Bang Observer (BBO), is probably at least twenty years away from being launched. It would likely have a design similar to LISA's, but with a smaller triangle whose sides will be a few hundred thousand miles across. The Big Bang Observer will be sensitive enough to detect the nearby sources individually. Once their contributions are identified and filtered from the data, the spacecraft array will also be sensitive enough to directly detect the much weaker cosmic gravitational waves streaming in from all directions, provided the inflationary picture is right. Because the Big Bang Observer will probe much shorter wavelengths than those detected using the B-mode polarization of the cosmic background radiation, it will provide a completely independent test that may settle the debate between the inflationary and the cyclic pictures.

The Moment of Truth

Just when the decisive moment in the sixth milestone test will come depends on a combination of unpredictable factors. The WMAP three-year results were the first real chance to detect cosmic gravitational waves, and they came up blank. At the same time, WMAP has demonstrated that the cosmos has not laid down an obstruction to finding a B-mode signal if gravitational waves account for at least 1 percent of the hot and cold spots in the WMAP image. This is enough to thoroughly test the predictions of the simplest inflationary models, which are even more strongly favored now that WMAP appears to have detected red tilt. There is a small chance that WMAP just missed detecting a signal, and that the Planck satellite or QUIET or other experiments planned for the next five years will be lucky and detect a signal. However, these experiments offer relatively modest improvements over the current limits. The technology for making a definitive measurement with a new satellite experiment is feasible, and the cost is reasonable compared with much grander projects now being considered by the National Aeronautics and Space Administration or the European Space Agency. Although a mission could be ready in five years, the recent redirection of NASA toward a manned mission to Mars has pushed the project off the U.S. agenda until at least 2018. ESA will be focusing on the Planck satellite in the short run. If the United States does not change its course, perhaps Europe,

Japan, or China will seize the opportunity to launch its own mission, recognizing the chance of making one of the great discoveries of the twenty-first century at comparatively little expense.

Successive efforts to detect cosmic gravitational waves, including the dedicated Inflation Probe satellite a decade from now, may fail. Then, the scales will tip heavily toward the cyclic picture, but a definitive resolution will take longer. The development of an improved gravitational wave detector, such as the Big Bang Observer, and other corroborative tests will probably take two or more decades to pursue. In the meantime, theorists will attempt to develop both cosmological pictures and continue to search for other ways in which the inflationary and cyclic models may be distinguished.

The only sure thing is that cosmologists will pursue the quest with every resource, skill, and technique available to them. Search they must, for the stakes have been raised. The competition between the inflationary and the cyclic models is no longer limited to the origin of galaxies and the nature of the big bang. The debate has expanded to include the future of fundamental physics and the nature of science itself.

chapter ten

Inflationary Multiverse or Cyclic Universe

Maybe nature is fundamentally ugly, chaotic and complicated.
But if it's like that, then I want out.

—Steven Weinberg

The previous chapters have focused on what the cyclic and inflationary models claim about the part of the universe that can be observed. Within this limited region of space, the models agree so closely that only highly refined measurements using the most advanced technologies can tell them apart. Where the two models sharply diverge is in their picture of what lies beyond the region of space we can see. In fact, their answers to many interesting questions—Does the universe evolve in the same way everywhere in space? Do the same physical laws apply throughout the universe? Are the large-scale properties of the universe random and unpredictable, or simple and regular?— could hardly differ more.

In the inflationary model, the vast majority of space is in a wild, uncontrolled state, undergoing violent, high-energy inflation. Scattered sparsely through this hostile expanse are regions where inflation has ended and radiation and matter have been produced, but these regions are of widely varying types. The conglomeration is not a universe at all but, rather, a *multiverse,* containing an infinite number of very different regions. According to this picture, the Earth is at a very unusual location, one of the very rare places in the multiverse where life is possible: what can or will be seen, even on the largest astronomical scales, is strongly constrained by what is needed for human existence.

The cyclic model represents a very different perspective, one in which the universe is almost the same everywhere. Every region of space undergoes controlled evolution through a series of regularly repeating cycles, which each start with a bang and end with a crunch, and in which dark energy plays a critical role in keeping the cycles on track. Every region makes galaxies, stars, planets, and, presumably, life itself, over and over again. Rather than the universe being a statistical fluke, it is the inevitable outcome of dynamical evolution governed by the laws of physics.

Although the differences between these models are not observationally testable (at least not in the foreseeable future), they are of much more than purely philosophical interest. They lead to diverging views of the future of cosmology and fundamental science itself. If the inflationary landscape model is to be believed, science has now reached the limits of what it can ever explain. Many basic properties of the universe, for example, the dark energy density, are just accidents and will never be predicted with any precision. Instead, they must be accepted as facts of our own peculiar situation in the "multiverse" of possible universes.

The cyclic hypothesis leads to a far more optimistic outlook for fundamental science. According to this view, the universe is a single, coherent entity that exists in a stable cycling state whose properties can eventually be understood as a consequence of the basic laws of nature.

This chapter contrasts the two very different perspectives. A dramatic illustration is what each has to say about the cosmological constant problem, widely regarded as one of the greatest challenges in science today. Neither the inflationary nor the cyclic models were designed to address the problem, so it is interesting to compare what kinds of solutions they suggest. The result is revealing. In one picture, the tiny value of the cosmological constant (or, equivalently, the dark energy density) is ascribed to chance and anthropic selection: the assertion that the presence of intelligent life plays an important role in determining the physical properties of the universe in the region of space we observe. In the other, the observed value may be explained as the consequence of very slow evolution that takes place over the course of many cycles.

"Eternal" Inflation and Guth's Rabbit

According to the inflationary model, the region of the universe observed today originally occupied an infinitesimal volume of primordial space. This little piece may have been highly warped and contorted at the start, but inflation blew it up to a vast size, smoothing out the whole expanse in the process. Inflation came to an end when the energy driving it decayed into matter and radiation, as described in chapter 3.

Also described in chapter 3 were the quantum fluctuations that cause the decay of inflationary energy to occur sooner in some places than others. The effect is due to the buildup of random quantum kicks every 10^{-35} seconds that slightly speed up or slow down the progress of inflation. On average, there are an equal number of kicks of each type, that speed up and slow down inflation, but the kicks do not balance out precisely everywhere. Instead, some locations in space will experience more kicks of one type and others will experience more kicks of the other type. The small net differences from place to place produce small temperature variations like those seen in the WMAP picture and small density variations like those needed to seed the formation of galaxies.

This, however, is not the whole story. The random kicks nearly average out in most places, but in some rare regions, an improbably long sequence of upward kicks causes inflation to last much longer than average. You might suppose that these improbable regions are so few and far between that they can be neglected. Not so! Inflation literally magnifies their importance. Whereas some regions terminated inflation and began to expand much more slowly, the improbable regions continued to inflate at an incredible rate, doubling in size every 10^{-35} seconds.

Within a few instants, the roles switch. The rare parts of inflating space blow up to vast size and quickly dwarf the regions like ours that have completed inflation and filled with a smooth distribution of matter and radiation. At each future instant of time, the rapid inflationary growth continues, continually spawning new pockets where inflation has ended. Once a pocket forms, it grows outward, eating away at the smooth, uniform inflationary energy from the inside. But this cannot outpace the rapid stretching of the space separating the pockets. In fact, the stretching is so fast that no light, particles, or information of any kind can be exchanged between pockets. However big the pockets grow, there always remain far larger regions in between that continue to inflate at an incredible pace. The result is called *eternal inflation*, an idea that emerged in the early 1980s in papers by Andrei Linde, now at Stanford University, by Alexander Vilenkin at Tufts University, and by Paul. The image it conjures is of a vast, rapidly stretching fabric of space marked with rare pockets containing matter, radiation, galaxies, and stars. Alan Guth, who first introduced the concept of inflation in 1980, refers to these separate regions that have terminated inflation as *pocket universes*.

The "eternal inflation" moniker is somewhat misleading. To many people, the word "eternal" implies an infinite past and future. But eternal inflation is only infinite to the future. If one traces the inflating fabric backward in time it shrinks away to a singularity just 13.7 billion years ago. Therefore the scenario of eternal inflation does not eliminate the need for a creation event. Proponents nevertheless claim inflation is so powerful that the outcome does not depend on the details of the primal event. But the only way to prove the claim would be to understand the creation event itself.

Some suggest that creation automatically leads to "chaotic inflation." This idea asserts, without proof, that the violent event that created space and time simultaneously generated a distribution of inflationary energy whose density varies randomly over space. Naturally, if the range is sufficiently broad, there is an excellent chance that the density is high enough somewhere that inflation takes hold. Once ignited, eternal inflation takes off. In this way, proponents argue that inflationary initial conditions are "generic." But this conclusion is hard to assess since it relies on an unproven assumption about some indefinite creation event.

Some theorists, such as James Hartle, Stephen Hawking, and Alexander Vilenkin, have tried to develop a quantum field theory description of how space and time might be created "from nothing" and to check if this leads to an inflationary universe. These attempts rely on quantum indeterminacy and inflationary expansion, not just to sustain inflation but to describe how a small new universe can fluctuate into existence in the first place. The subject has been controversial because there are no rigorous physical principles that dictate how to go from "nothing" to "something." Thus far, the most reliable approximations suggest that quantum creation results in an empty universe rather than one undergoing high-energy inflation. And the introduction of string theory and M theory have not changed the story. So, at present, the inflationary model is incomplete because there is no compelling reason for the universe to emerge in an inflating state.

Many proponents of inflation consider the ability of inflation to multiply and sustain itself to be a vital aspect of the theory. For example, although Guth had not realized the eternal nature of inflation when he initially proposed the idea, he has become a leading advocate of its importance. "I would argue that once one accepts eternal inflation as a logical possibility," he has written, "then there is no contest in comparing an eternally inflating version of inflation with any theory that is not eternal. Consider the analogy of going into the woods and finding some rare species of rabbit that has never before been seen. You could either assume that the rabbit was created by a unique cosmic event involving the improbable collision of a huge number of molecules, or, you could assume that the rabbit was the result of the normal process of rapid reproduction, even though there are no visible candidates for the rabbit's parents. I think we would all consider the latter possibility to be far more plausible. Once we become convinced that universes can eternally reproduce, then the situation becomes very similar, and the same logic would apply. It seems far more plausible that our universe was the result of universe reproduction than that it was created by a unique cosmic event."

Guth's rabbit analogy argues for a cosmological model capable of producing many regions like the one occupied by the Earth. However, in the context of inflation, the analogy is a little misleading. In eternal inflation, the habitable pocket universe regions are not what self-reproduce, at least not directly. Rather, the precursor, rapidly inflating regions, are the ones that reproduce themselves like mad. Eternal inflation is like a giant runaway engine creating unlimited amounts of space filled with nothing but inflationary energy. For every habitable pocket, there is a vastly greater expanse of uninhabitable, inflating space. Is it reasonable to suppose that nature be so profligate, in order to create the one region that is actually observed?

The Inflationary Multiverse: Lost in Space

So far, eternal inflation has been described as if every pocket universe it creates is exactly like ours. But in recent years theorists have realized that the eternal inflation picture leads to an even more extravagant possibility, known as the "inflationary multiverse," in which the physical conditions vary greatly from pocket to pocket.

The basic reason is that just as the energy density and gravitational waves undergo quantum jitters during inflation, so do all other quantities. Quantum jumps can produce great variations in the properties of the pocket universes. For example, some undergo considerably less inflation and hence end up considerably curved or warped. Since inflation is supposed to explain why the universe is smooth and flat, one might hope that such mutations are rare. But can this be proved?

One approach would be to follow a patch of inflating universe for a long time and then take a census of the pocket universes that form within it. From the census, one could determine the fraction of space occupied by pocket universes of each different type. However, there is a serious problem with this idea. The calculation relies on being able to make your evaluation at a common "time" all over space, a kind of synchronization. But a key principle of general relativity is that there is no unique way of synchronizing clocks that are spread over space. In this case, making different rules for synchronizing clocks rearranges the time order in which pockets are created and completely changes the results of the census. One choice suggests that almost all pockets are flat. Another suggests that almost all are highly curved. And there are synchronization choices that give every answer in between. According to Einstein's theory, all choices are equally valid.

Guth has compared the problem to that of determining what proportion of numbers are even or odd. The answer might seem obvious at first: order the numbers as pairs of consecutive numbers—(1,2), (3,4), (5,6), (7,8), and so on—for example, and there seem to be equal numbers of odd and even integers, since each new pair brings one even and one odd number. This is analogous to choosing a particular synchronization of time for taking the census of pocket universes. But why not order them in triplets so that each new triplet brings one odd and two even numbers: (1,2,4), (3,6,8), (5,10,12), (7,14,16), and so on. In this ordering, you might conclude that for every odd integer there are twice as many even integers. One can easily come up with alternative orderings and obtain any answer at all. For a finite set of numbers, there is a unique solution that can be found simply by counting. But for an infinite set, there is no well-defined answer. Similarly, there is no way of telling what proportion of space is occupied by each type of pocket universe.

Eternal Inflation and the String Landscape

Now consider bringing string theory into this picture. String theory's remarkable successes have been recounted in previous chapters. It is designed to be a consistent quantum theory of gravity and a natural unification of gravity with the other forces of nature. As discussed in chapter 6, though, string theory proposes that all of the basic physical properties of the world—the forces, the particles, the dark energy density, and so on—are determined by the size and shape of the extra dimensions. To understand why all the forces and particles are arranged as they are, one needs to know the structure of the extra dimensions as the universe emerged from the big bang. One might hope that the powerful combination of string theory and inflationary cosmology would dictate the answer, uniquely specifying the fundamental forces and particles and resolving the ambiguities of eternal inflation.

It is possible, though, that string theory only exacerbates the situation. Chapter 6 described how string theorists have been studying mechanisms for fixing the sizes and shapes of extra dimensions by twisting branes and fields around them. In order to compare all the options, it is helpful to picture an *energy landscape*. Every point on the landscape corresponds to some choice for the sizes and shapes of the extra dimensions. Moving across the landscape corresponds to changing the sizes or shapes or both. The altitude at each point represents the potential energy density if space has the particular choice of sizes and shapes of the extra dimensions. The wild up-and-down landscape means that the potential energy density is highly sensitive to the configuration of the extra dimensions.

If a region of space starts out with large vacuum energy, corresponding to a mountain on the energy landscape, the extra dimensions will tend to adjust themselves to move toward a lower-energy state. The path they follow is just like that of a ball rolling on a hilly surface: eventually, like the ball, they will settle in a hollow or trough in the energy landscape.

If there were just one low point in the energy landscape, with the terrain rising all around it, then, no matter how they started, the extra dimensions would always slide down to this unique minimum. Recent calculations in string theory, although not conclusive, suggest a much more complex landscape, more like a Himalayan mountain range with jagged peaks and clefts. Scattered throughout are little hollows, numbering at least 10^{1000} and probably more, in which the extra dimensions can get stuck for a very long time.

If the extra dimensions start out on a high plateau, they can provide the inflationary energy to drive a powerful burst of inflationary expansion as they roll down to a low-energy state. As they do so, their motion is strongly influenced by quantum jitter. Which particular hollow they end up in is determined by chance, and that is what determines the properties of the pocket universe that then forms. Different pockets will have different laws of physics, different types of matter and energy, and different concentrations of ordinary matter, dark matter, and dark energy within them. With so many hollows to choose from, the range of possibilities is incalculably vast. And there is no way, even in principle, to assess the proportion of space that ends up in pockets of each type.

Throughout the twentieth century, fundamental physics has been driven by the vision of Einstein and others that the universe should be simple and comprehensible. Particle physics experiments seemed to confirm this view by showing that the laws of fundamental physics unify and become more symmetrical as temperature and energy increase. Similarly, astronomical observations showed that the universe—at least the part that can be seen—emerged from the big bang in an extraordinarily smooth, simple state. However, according to the inflationary landscape picture, all of this apparent simplicity is just an illusion. There is nothing unique about the laws of physics, and almost any laws are possible. The universe appears smooth and uniform because astronomers can see only a tiny patch of it: its true, wild, random structure on ultralarge scales is unobservable. All of the physical properties of the observable universe are essentially an accident whose history can never be unraveled. Instead of Einstein's dream, the universe is Einstein's worst nightmare.

The Anthropic Principle to the Rescue?

This turn of events was shocking. The marriage of two powerfully predictive concepts, string theory and inflationary cosmology, produced an inflationary landscape–multiverse picture that is neither predictive nor verifiable. Yet lurking in the wings was an old, familiar character, which appeared to some to be just what was needed to snatch victory from the jaws of defeat.

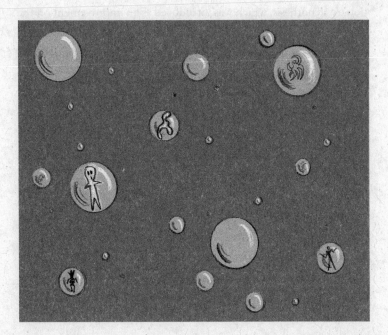

The *anthropic principle* is a widespread notion that, under different names, has been invoked in various scientific contexts over the centuries. Its current name is due to Brandon Carter from Observatoire de Paris–Meudon, who introduced the term in 1973. To some, the term means only that the physical laws and conditions that govern the universe must be compatible with the fact that life exists. In this form, the principle is not contentious. The fundamental laws that govern the universe must be consistent with all observations, including the fact that part of it is habitable.

The controversy begins when the anthropic principle is used to explain what appear to be fine-tuned properties of the observable universe that cannot be explained by current theories. Historically, scientists consider fine-tuning to be a sign that a current theory is flawed or incomplete, and they view fine-tunings as profound hints about how to improve their theories. A relevant example is the inflationary hypothesis, which was introduced as an improvement on the original big bang theory to explain the homogeneity and flatness of the universe. The anthropic principle, though, suggests a different approach: Do not change the current theory just because it has finely tuned properties. Instead, imagine that there is a multiverse in which those properties vary randomly from one universe to the next and that, in most, the universes have no finely tuned properties. But the Earth lies in a rare universe out of the multiverse of possibilities because fine-tuning is a prerequisite for life to evolve. In other words, instead of the physical laws explaining the complexity of life, life is invoked to explain the complexity of the physical laws.

Before the inflationary landscape picture, anthropic selection was generally regarded as a sort of scientific parlor game not to be played in serious company. Its proponents would assume that we live in a multiverse containing separate universes with a wide range of physical properties, and then they would try to show that life would be most probable in the universe with the properties we actually observe. But there were no clear rules to the game, and there were no independent reasons to believe that any of the counterfactual universes actually exist. The inflationary landscape picture gave the anthropic principle a new lease on life by suggesting that the multiverse might naturally arise from string theory and that inflation could naturally populate the landscape of possibilities. Suddenly, the parlor game seemed to gain sound scientific footing.

The simplest example of a quantity that might be "anthropically selected" is the density of dark energy. In the energy landscape, this corresponds to the height of the particular hollow into which the extra dimensions have settled. Since each pocket universe settles in a different hollow, there are a huge number of possible values for the dark energy density. Most of these values are too large to allow life to form. If the dark energy is too positive its repulsive gravity will blow up the universe and dilute the matter away before galaxies can form. If it is too negative, the universe will collapse too soon for life to have evolved. But given such a dense set of possible values, a few pocket universes will have a dark energy density small enough to allow galaxies and life to form. By the same reasoning, even fewer will have a much smaller dark energy density. Therefore, in pockets where there is life, the expectation is that the dark energy density will be just small enough to allow galaxies to form, but no smaller. Similar remarks apply to other parameters that might be anthropically selected: the number of dimensions, the ratios of force strengths and particle masses, and so on. In each case, one can argue that life would be difficult if the world were not the way it is. And, just like the argument for the dark energy density, if the anthropic selection is at work, the expectation is that conditions should be barely compatible with life. By adding the anthropic principle to the mix in this way, proponents claim, the inflationary landscape picture can be transformed into a predictive and testable theory.

The anthropic principle has had a major impact on recent cosmological thinking. Some highly respected theorists have begun writing papers and books and giving talks that invoke anthropic reasoning to address fundamental problems in cosmology, providing the concept with increased visibility and perhaps even credence among the general public. At the same time, a substantial fraction of the scientific community, perhaps a majority, is opposed to the anthropic principle on the grounds that it is nonscientific.

The two of us confess to being among those who are skeptical about the anthropic principle as a panacea for fundamental physics. The main problem is that it relies on a host of assumptions that cannot possibly be tested: the existence of the multiverse; the notion that different pockets have different physical laws; the idea that the physical conditions in each pocket are chosen by chance, although the probability distribution cannot be calculated; various suppositions about which physical properties vary from pocket to pocket and which do not; further assumptions about the conditions necessary for life to exist; and so on. There is a tremendous amount of arbitrariness in choosing which selection principles ought to be imposed, as illustrated by the number of disparate anthropic models in the literature.

Many scientists, including the two of us, feel it is important to insist that science should remain based on the principle that statements have meaning only if they can be verified or refuted. Ideas whose assumptions can never be tested lie outside the realm of science. To be sure, proponents are quick to argue that the predictions of the anthropic principle can be tested even if the assumptions cannot. For example, anthropic arguments are used to give rough predictions for the dark energy density, and astronomers can make measurements to check those predictions. But there is a serious logical flaw in this argument. Just because a prediction is consistent with the evidence does not mean the theory is right. One must show that the theory has correctly identified the root cause of the phenomenon.

When leading physicists suggest that the anthropic principle provides a legitimate answer to deep puzzles, it discourages theorists, especially young ones, from pursuing fresh attacks on the problems. It creates an unfair competition. Normally, a scientific idea must survive the rigorous tests of mathematical consistency and experimental verification. This often requires years of study, development, and modification, with the constant risk that the idea might fail. In fact, many more ideas fail than succeed. The anthropic principle, with its malleability and reliance on untestable assumptions, is never at risk of being proved wrong. On this uneven playing field, few will feel encouraged to seek scientifically refutable alternatives.

For this to be happening in cosmology at this particular moment is profoundly disturbing. After thousands of years of speculating about what the universe might be like, humanity has only recently developed the technology to see what is actually there. The picture turns out to be astonishingly simple and increasingly uniform the further out one looks. The current understanding of string theory, the best hope for explaining this simplicity, is only in its infancy but already appears to contain many of the needed ingredients. It seems much too early to turn from ordinary science to the anthropic principle.

Another concern is the impact of anthropic reasoning on science and society generally. By invoking anthropic reasoning, cosmologists blur the historic distinction between physics and metaphysics. It is not possible to draw a clean line between the anthropic principle, with its reliance on untestable assumptions, and other untestable beliefs and superstitions. The long-term effect of basing theories on anthropic reasoning could be to undermine the role of science in enlightening humankind, in insisting on objectivity and proof, and in steering society away from poor decisions based on myths and fallacies.

Lost in Time

The inflationary model also says something about time and the flow of cosmic history within our pocket universe. The basic picture is a series of one-of-a-kind events. The universe starts with a creation event, followed shortly by inflation. When inflation ends, there is a hot radiation era and a matter-dominated era lasting 9 billion years or so. Finally, dark energy takes over, stretching space out and diluting away the matter and radiation into a nearly pristine vacuum.

The universe as observed today, teeming with stars and galaxies, is a brief hiccup in cosmic history, sandwiched between two periods of accelerated expansion, which are driven by inflationary energy at one end and dark energy at the other. In between, the sparks created by the end of inflation evolve in the most marvelous way to form atoms, then molecules, then dust, and onward to the fascinating and diverse universe that surrounds us all. One marvels at the complexity that has emerged from such simple origins. The ever larger cosmic structures that have formed are impressive: the diverse population of galaxies, clusters of galaxies, superclusters, giant walls of galaxies stretched across the sky, and bubble-like voids.

According to the inflationary model, though, this steady progression has now reached an end. Now that dark energy has overtaken the universe, the formation of yet larger structures cannot take place. In a trillion years, the size of the universe will double over a hundred times. Matter will be so thinly spread out that no new stars or structures can form. The luminous universe observed today will become an empty, cold, structureless, lifeless vacuum that will last forever. (Quantum jitter can transform some of the wasteland back into an inflating patch and new regions of matter, but the fluctuations occur far too rarely to significantly affect the picture we have painted.)

In the inflationary model, most of cosmic history is spent creating more and more vacuum. Only for the blink of an eye does a pocket universe take advantage of the complexity and vitality of the underlying physical laws, by forming an oasis of matter, radiation, and galaxies. But just a few cosmic instants after all this has come to fruition, the oasis is snuffed out. True, in eternal inflation, pocket universes continue to form within the inflating state. A few will allow all of the complexity observed today. But each pocket universe shares the same fate: to be rapidly extinguished soon after it is created, and replaced with a featureless void that will last virtually forever.

The Inflationary Schism?

Something strange has happened to the inflationary picture. Considerations of eternal inflation, the multiverse, and the string landscape have transformed the inflationary picture in a way that belies its popular image. What began as an efficient mechanism for explaining the apparent simplicity of the universe—that is, the smoothness and uniformity of the observable universe—has been turned into a runaway process that seems extraordinarily wasteful and unpredictable in its use of space and time.

The root cause is that inflation has proven to be too powerful to control. It is able to use gravity to wrest control over space. Then this power is utilized to create yet more inflating space. Infinite numbers of pocket universes of all types are spawned as a by-product, but inflation operates at such high energies that it generates them all indiscriminately. Regardless of the low-energy laws of physics and the final state of the universe, inflation populates all possibilities to an infinite degree, making it impossible to explain why the observable universe has the particular physical properties it has. One is left to conclude that the large-scale uniformity of the universe and the simple physical laws that appear to govern everything that has ever been observed are actually atypical of the universe as a whole, rare accidents that exist only because of random chance.

This raises a provocative question: Is the inflationary model, as it currently stands, a valid scientific theory? Or is the discovery of infinite possibilities a sign of a serious flaw, both logically and scientifically? A poll of cosmologists today would probably reveal that an overwhelming majority believe inflation to be not only a valid idea but the most likely explanation for the universe we observe. But if the poll went further to ask whether the infinite number of possible universes is a problem, it would probably reveal a split into very different camps. One camp loudly extols the idea of multiverse, landscape, and infinite possibilities. For example, Steven Weinberg writes, "Most advances in the history of science have been marked by discoveries about nature, but at certain turning points we have made discoveries about science itself. These discoveries lead to changes in how we score our work, in what we consider to be an acceptable theory. . . . Now we may be at a new turning point, a radical change in what we accept as a legitimate foundation for a physical theory."

Meanwhile the other camp is quietly worried. Its members latched onto inflation when the idea seemed to lead to a simple and unique outcome—or at least that's the way it was taught to them. They are encouraged by the fact that the observations agree with the simple predictions and hope that the problem of an infinite number of universes will be solved sometime in the future. There are historical precedents to support their hope. For example, in the first attempts to construct quantum field theories that unify electromagnetic and weak forces, theorists found that the mathematical equations were "nonrenormalizable," meaning that they had an infinite number of parameters that could be adjusted independently. Each choice of parameters would give distinct predictions. To decide which choice describes the real world would require an infinite number of independent measurements. Some physicists concluded that the infinite possibilities were inevitable and gave up on quantum field theory as a useful tool. Others viewed nonrenormalizability as a serious flaw but a potentially curable one. Over the next decades, they took on the uncertain search for an improved theory, and they eventually succeeded in understanding how to obtain a renormalizable theory with a finite number of parameters. Steven Weinberg, quoted earlier, was one of the codiscoverers of that successful theory.

The same could happen with inflation. Perhaps the inflationary model, as it currently stands, is incomplete and a mechanism can be added that will tame its wild, runaway behavior so that a unique outcome emerges that is consistent with the real universe. Most critics would be prepared to accept the inflationary picture should this occur.

But in the interim—even now—a schism may be emerging. Many outstanding leaders in cosmology, astrophysics, and string theory, including Andrei Linde, Martin Rees, and Leonard Susskind, have come to believe that the uncontrollable features are essential, to be celebrated rather than tamed. Others, like David Gross, hold firm to the conviction that the current situation is unacceptable and that a better theory must be possible. There is a real conflict developing between the two points of view. For example, after Gross, quoting Winston Churchill, exhorted his colleagues, "Never, never, never, never give up!," Susskind retorted, "But the field of physics is littered with the corpses of stubborn old men who didn't know when to give up." In his opening address at a meeting entitled Expectations of a Final Theory, Weinberg offered an intermediate perspective: "I noticed for sale the October issue of a magazine called *Astronomy,* having on the cover the headline 'Why You Live in Multiple Universes.' Inside I found a report of a discussion at Stanford at which Martin Rees said that he was sufficiently confident about the multiverse to bet his dog's life on it, while Andrei Linde said he would bet his own life. As for me, I have just enough confidence about the multiverse to bet the lives of both Andrei Linde *and* Martin Rees's dog." The repartee is chosen to amuse, but behind it lies a serious statement about differing visions of what is and is not valid science and how close scientists are to a final theory today.

There is the possibility that both sides are right. In the end, there may be no way to avoid the uncontrollable nature of inflation. At the same time, as a broader range of scientists become aware of this characteristic, they may not give up. They may, instead, reject the inflationary landscape approach altogether in favor of an alternative with more explanatory and predictive power.

The Parsimonious Universe

The cyclic model of the universe offers an alternative: a return to economy and predictability of the type envisioned by Einstein and others. The underlying mechanism driving the cycles is gentle and self-regulating. The collisions between the two branes occur at modest speeds (well below the speed of light). The dark energy density is always low, and there is no runaway to high-energy states where vast expanses of inflationary universe are created. Instead, dark energy acts as a shock absorber that keeps the cycles under control, suppressing the effects of random fluctuations so that the regular, periodic evolution is kept on track.

In the cyclic model almost all of the matter and radiation observed today was created at the last bang, and its subsequent evolution has been more or less identical to that predicted by inflation. Matter and radiation created at the previous bang were diluted away during the dark energy phase of the last cycle and have an unobservable low density today. Hence, despite the fact that the cyclic universe is far older than 14 billion years, it makes almost the same predictions as the inflationary model for the region of space that astronomers observe.

Yet the cyclic and inflationary models are completely different on large scales. In the inflationary picture, regions of space with galaxies and stars occur only in rare pockets that are separated from one another by unimaginably vast expanses of empty space. By contrast, the cyclic model predicts that everywhere in space has a distribution of galaxies and stars similar to what is seen from the Earth. Occasionally, rare quantum jumps or other events will create black holes whose strong gravitational field will disrupt the smooth cycling in their vicinity. However, the regular stretching of the branes (by a factor of 10^{50} or more during each cycle) ensures that the black holes remain isolated and do not disrupt the cycling taking place everywhere else.

Because the cyclic model is a self-reproducing cosmology, it satisfies Guth's criterion as expressed through his playful rabbit analogy. In fact, the analogy is far more apt for the cyclic model than for inflation: in the cyclic model, the habitable regions are the ones that "breed like rabbits," multiplying exponentially in number with every new cycle.

Simplicity and parsimony reign. Virtually every patch produces galaxies, stars, planets, and life, over and over and over again. Through controlled recycling, Einstein's vision of a universe that is smooth and predictable on very large scales emerges naturally. The more one learns about string theory, the more one can understand the physical principles underlying the controls. In particular, even if the string energy landscape is as complex as imagined, long-term cycling is possible only under special conditions. Thus the universe, or the vast majority of it, may be naturally driven toward an energy valley with the properties that astronomers and particle physicists observe today.

Inflation Versus the Cyclic Model

The inflationary model is older, more studied, and more widely known than the cyclic model. Many cosmologists already accept it as the likely explanation for the flatness, uniformity, and large-scale structure of the universe. To a large degree, the acceptance is based on the simpler picture of inflation that emerged twenty-five years ago, before there was any notion of a multiverse and before the anthropic principle became associated with the model. As the scientific community takes a closer look at the modern version of inflation and compares it to the new alternative, the cyclic universe, several distinctive features of the cyclic model will become more widely appreciated:

Fewer Ingredients: Inflation requires both inflationary energy to drive the period of high-energy acceleration in the early universe and some form of dark energy to explain the current low-energy acceleration. The cyclic model requires only dark energy.

More Parsimonious: Inflation creates unbounded amounts of space and time in order to create a few rare pockets that look like what is actually observed. Those rare regions last 10 billion years or so before being permanently overcome by dark energy. In the cyclic model, almost all of space has the same physical laws and the same structure. Galaxies, planets, and life are ubiquitous, and each region of space produces them over and over again.

More Selective: Inflation is insensitive to the physical conditions in the universe after inflation is completed. So, if there is an energy landscape, inflation creates pocket universes of every possible minimum with every possible set of physical properties. Only an infinitesimal minority of pockets are habitable, so the existence of humanity relies upon a rare event in a random process. Cosmology is forced to turn to the anthropic principle to explain the state of the universe. By contrast, in the cyclic model, only regions with certain select properties self-reproduce, and these regions eventually occupy nearly all of space. The universe can, in principle, be explained by ordinary science; that is, the laws of fundamental physics will themselves drive the universe into a cyclically self-reproducing state, whose physical properties should be calculable someday and shown to match what is actually observed.

These features alone should be appealing enough to earn the upstart cyclic model serious consideration. But a hypothesis may be wrong even if it has features that one finds philosophically attractive. Passing the next few rounds of experimental tests is essential and would add further support to the cyclic model. Yet what often wins the day for a new theoretical idea is the discovery that it opens a new avenue for resolving long-standing puzzles that it was never intended to address. With this thought in mind, consider one more feature of the cyclic picture, which may prove to be of critical importance in explaining the present state of the universe.

More Time

The most obvious property of a cyclic universe is that it is very old. Every patch of space, including the volume now visible to us, has existed for far longer than 14 billion years, the age according to the inflationary picture. Each cycle lasts about a trillion years, and there is no known limit to how many cycles there could have been in the past. Perhaps the number is infinite. Or maybe there was some "beginning" in the distant past, after which the universe was driven toward regular cycling behavior by its natural stabilizing properties. Either way, including prior cycles, the cyclic universe is exponentially older than the conventional inflationary universe.

Having more time allows for new solutions to some of the most difficult cosmological problems—not just metaphysical issues regarding space beyond the limits of detection but also key mysteries concerning the physical properties that can be observed, which neither the inflationary nor the cyclic model was designed to address. The thorniest and most well-known is the *cosmological constant problem*, which many physicists regard as the most profound enigma in all of science.

When Einstein first introduced the cosmological constant in his seminal 1917 paper, he had no idea how much trouble he was creating. His motivation, as explained previously, was to construct a static model of the universe. After Hubble found that the universe is expanding, Einstein thought he could simply discard the idea, later disowning it as his "greatest blunder." However, others later realized that the cosmological constant is actually the vacuum energy, a quantity that should, in principle, be calculable from the fundamental laws of physics. According to those laws, empty space is not the boring place you might imagine. On microscopic scales, it is full of activity because quantum processes are constantly creating and then annihilating particles and antiparticles of all types. These particle-antiparticle pairs each contribute to the vacuum energy. Knowing all the particles and forces, the contributions can be estimated and added together, although the result depends strongly on the behavior of the theory at extremely high energies, where little is known from experiment. Nevertheless, if one naively extrapolates the known forces and particles to high energies, the calculated vacuum energy is an incredible 10^{120} times larger than the measured value.

For a time, physicists thought the problem might be resolved by some yet-to-be-discovered principle of physics that cancels out the cosmological constant at the moment the universe emerges from the big bang. This idea fails because certain contributions to the cosmological constant do not kick in until later, as the universe cools down. The ideal solution is, therefore, a "relaxation mechanism" that causes the cosmological constant to change in response to any new contributions. Conceivably, each contribution could be canceled once it arises, provided the relaxation mechanism is fast enough to prevent the cosmological constant from dominating until very late in the day. However, since inflationary energy behaves just like a cosmological constant in causing the expansion of the universe to speed up, any efficient mechanism for relaxing the cosmological constant would also be likely to prevent inflation from occurring.

This imposes new restrictions. The mechanism must not work so efficiently during inflation that it prevents the accelerated expansion required to make the universe smooth and flat. After inflation, the canceling mechanism must be extremely efficient or else the cosmological constant will dominate over the matter or radiation and galaxies cannot form. Then, after 10 billion years, the canceling mechanism must turn off or slow down so that dark energy can dominate. In other words, for a relaxation mechanism to fit within the current inflationary model, it must be slow, then fast, and then slow once again. A mechanism that must turn somersaults like this sounds rather implausible.

By dashing the best hope for a solution, the discovery of dark energy caused many theoretical physicists to turn to the anthropic principle out of despair. Essentially, they conceded the battle. Having failed to find a mechanism for making the cosmological constant small, they concluded that it must be large in most places, and that the observed universe must be an anomaly.

As we were inventing the cyclic model, we wondered if it might offer a new approach for resolving certain cosmological puzzles that appear intractable within the conventional picture. For example, in the inflationary picture, it seems a complete fluke that when the universe first comes into existence, it is already set to have a tiny density of dark matter that becomes important only after 75,000 years and a tiny cosmological constant that becomes significant only after 9 billion years. How can forces that act only in the opening instants of the universe be designed to fix the conditions "just right" at a much, much later time? Some "advanced knowledge" of where the universe is headed seems to be required. Within a cyclic universe, though, that "advanced knowledge" could exist, in principle, because the universe has already gone through many epochs before the big bang in which dark matter and dark energy have dominated.

But only as we were putting the finishing touches on this book did we come up with a physical process for implementing this idea, a natural relaxation mechanism in place of the anthropic principle. In this new approach, instead of a mechanism that is slow-fast-slow, the relaxation is always excruciatingly slow—so slow that the cosmological constant undergoes almost no change over the course of 14 billion years or even over the course of a single cycle. Slow-fast-slow is unnecessary because there is no period of high-energy inflation in the cyclic model. And "ultraslow" is conceivable because the universe, having undergone countless cycles before the big bang, is much older than 14 billion years. An ultraslow relaxation process would be useless in the conventional big bang picture because the universe would be completely empty by the time the cosmological constant reached the small value observed today. In the cyclic model, though, fresh matter and radiation are created during each cycle, so the galaxies and stars are regularly replenished even as the cosmological constant slowly decays.

Another important feature of the cyclic model is that while the extra dimension separating the branes expands and contracts on a regular basis, the branes themselves undergo a large net expansion in the course of each cycle. As a result, quantities that depend on the expansion and contraction of the extra dimension, such as the temperature and energy density after each big bang, are reset to the same value from one cycle to the next. But quantities that depend on properties of the branes alone, such as the total volume of three-dimensional space or the total entropy, can steadily increase (or decrease) from cycle to cycle. This is how Tolman's entropy problem is resolved, as described in chapter 8. Similarly, by having a relaxation mechanism that depends on the branes alone, it is possible for the cosmological constant to decrease steadily from cycle to cycle while the temperature and energy density are restored to the same value after each bang.

Of course, there has to be a plausible ultraslow mechanism that can progressively cancel the cosmological constant on the branes.

This requirement turns out to be simple to meet. Over the last few decades, in attempting to solve the cosmological constant problem within the context of inflationary models, theorists had already discovered several possibilities. These were rejected at the time because they were far *too slow*, so that the universe was empty by the time the cosmological constant reached the observed value. But, in the cyclic model, *too slow* is just fine.

A particularly appealing approach is one proposed over twenty years ago by Laurence Abbott from Columbia University. Abbott introduced a mechanism in which the cosmological constant can begin with a large value and creep its way downward through a series of tiny quantum jumps. The mechanism, which involves a quantum field that tunnels through a series of tiny energy barriers to states with lower and lower vacuum energy, requires exponentially increasing amounts of time as the cosmological constant approaches zero. As a result, the universe spends far more time at the last step above zero, when the cosmological constant is tiny, than it does on all the other steps combined.

Once settled on the last positive step, the universe will occasionally undergo a final quantum jump that changes the cosmological constant in a small volume of space to a negative value. These regions collapse under gravity in a very short time and have no long-range effect on the rest of the universe. Hence, almost all of space and time is spent in a state with a small, positive cosmological constant consistent with the dark energy observed today.

Abbott tried to apply his idea to an inflationary universe, but he abandoned it because the jumping process takes far too long. Here is where the cyclic model comes to the rescue. Abbott's idea can be incorporated into the cyclic model by having the quantum field driving his mechanism reside on the hidden brane, which, from a technical point of view, turns out to be a natural possibility. This means its behavior depends only on what happens to the brane itself, and not the regular expansion and contraction of the extra dimension.

The cosmological constant measured by astronomers is the total vacuum energy, a sum of contributions from both branes, as well as the potential energy associated with the springlike force causing the branes to collide every trillion years or so. Some of these contributions never change. Some, like the potential energy, return to the same value after each big bang. But for Abbott's quantum field, the contribution to the cosmological constant steadily decreases over time, without being disturbed by the cycling. As a result, the total vacuum energy, or cosmological contant, reached in the dark energy phase of each cycle also relaxes over the course of many cycles.

In this cyclic picture, two timescales characterize the evolution of the universe: the duration of a cycle, which lasts about a trillion years, and the time between downward jumps of the cosmological constant, which takes far longer as the universe approaches the bottom step. At each step downward, the branes undergo a rapidly increasing number of cycles. During the early stages when the cosmological constant is large, dark energy overtakes the universe at a much earlier point in each cycle than it does when the cosmological constant is as small as it is today. As a result, for these early cycles, matter never has the chance to dominate the universe and galaxies never form. However, the overwhelming majority of the cycles occur when the cosmological constant reaches the smallest positive step, the value it has today, and then galaxies form during each and every cycle. Our calculations showed that a patch of the universe would survive at least $10^{10^{100}}$ cycles at this last step.

The concept is surprisingly simple: the cosmological constant is much smaller than we expected because the universe is much older than we thought. This is possible because the universe is cyclic, which not only provides more time but also replenishes the universe with new matter and radiation, so that it is not empty by the time the cosmological constant becomes small.

It should be emphasized that we did not invent the cyclic model with the intent of solving the cosmological constant problem. Quite the contrary; our original idea had been that all cycles have the same broad properties, including the same value of the cosmological constant. In the absence of any other physical explanation, the value of the cosmological constant was artificially tuned to be in accord with what is observed, just as in the inflationary model. We had not thought about taking advantage of all the extra time the model naturally provided. With twenty-twenty hindsight, the ultraslow relaxation idea now seems obvious.

The contrast with the anthropic multiverse picture is stark. The multiverse picture relies on a series of untestable assumptions about parts of the universe that can never be observed: the existence of multiple universes, the variation of certain physical properties from universe to universe, the dependence of life on those physical properties, and so on. Also, the anthropic picture assumes that most of space is forever uninhabitable, with physical conditions that are never like those observed by astronomers and physicists, despite the fact that the laws and physical conditions are the same as far as can be seen. The cyclic solution does not depend on any of these assumptions. Instead, almost every patch of space anywhere in the universe evolves to the point where it is habitable and has the same properties that we observe—the same kinds of matter and energy, the same cosmic background radiation temperature, and the same kinds of structures. Most people will agree that, all else being equal, a cosmological model that predicts the same physical conditions we observe occur almost *everywhere* is vastly preferable to one that predicts those conditions occur almost *nowhere*. This being the case, we should not settle for a model of the *second kind* until very strong arguments have been made against all models of the *first kind*.

The cyclic model is a theory of the first kind. What are the main arguments against it? Certainly, the model is quite young and still incomplete. A key ingredient, the springlike force that draws the two branes together at regular intervals, is assumed to exist, but ultimately it must be established that the force arises naturally from M theory. Second, the mathematics describing what happens when two branes collide is not yet well-developed. One cannot yet be sure that branes bounce in the manner assumed in the cyclic model, although, as we mentioned in Chapter 7, some promising results in this direction have been obtained. Finally, there remain some questions about how the quantum-induced ripples produced on the two branes turn into scale-invariant density variations after the bang. A deeper understanding of the principles of M theory will likely be required before these questions are finally settled.

The inflationary model has had a twenty-year head start but it is nevertheless far from complete. The source of the inflationary energy is unknown and, so far, no convincing candidate has been identified in either string theory or M theory. In spite of several valiant attempts, inflationary theorists have so far been unable to explain how space and time began and why the big bang led to inflation in the first place. Finally, proponents have yet to come to grips with the runaway nature of inflation and the infinite possibilities in the landscape. Some seem prepared to accept these features and adopt anthropic reasoning, if necessary, to explain why, in a universe that explores all possibilities, the part of the universe we observe has the properties it does. In fact, some go so far as to advocate giving up on ever finding a theory of the first kind. Others, including us, consider it premature to concede defeat at a time when our understanding of both fundamental physics and cosmology is still in its formative stages. For those who share this perspective, the inflationary model must be regarded as seriously flawed unless new elements can be found that can tame inflation's wild behavior and transform the landscape notion into a powerfully predictive concept.

The next decade or two is likely to have an historic impact in determining which kind of theory and which specific model best explain the universe in which we live. Theoretical tools are already being developed that may make it possible to explore what actually happened at the big bang, and these could be decisive. Equally if not more important, there will be a host of new measurements, both laboratory experiments and astronomical observations, that will provide important new clues. We are optimistic that, through the collective efforts of experimentalists, observers, and theorists, the crucial breakthroughs will be made that will ultimately decide the debate.

Back to the Future

What will the future bring? If the inflationary model is correct, all of us live on a planet that is lost in the multiverse. Almost nowhere are the physical conditions like those we observe. And our rare pocket of the universe is running out of time. Dark energy has already overtaken all other forms of matter and radiation, and has taken command of the expansion of the universe.

In a trillion years, our home will be well on its way toward a vacuous oblivion. Virtually all the galaxies we see today will still exist, but the stars will be gradually burning out. There will in all likelihood still be stars, planets, and life. But the accelerated expansion due to dark energy will have spread out the galaxies so much that nothing beyond the Andromeda Galaxy will be visible to us.

The surviving civilizations in the Milky Way will know from the historical record that the universe was once filled with billions of galaxies, which emerged from tiny fluctuations in a hot plasma uniformly spread over space. But all the observational clues available today will be long gone by then. It is hard to imagine that a newly emerging civilization could piece together cosmic history on its own. In the inflationary model, therefore, the present is a unique epoch in the evolution of the universe where we can see both substantial amounts of the matter and radiation that dominated our past and the dark energy that will dominate our future. At other epochs, only one or the other would be detectable.

The same trillion-year prospectus applies if the cyclic model is correct, but it holds nearly everywhere in space, not just in isolated pockets. After a trillion years, however, the story changes dramatically. The branes begin to approach each other, the dark energy decreases, and expansion slowly grinds to a halt. There will be no galaxies or other distant sources that future observers can use to detect the expansion rate, unless the future civilizations send regular test probes beyond the Milky Way. Yet there will be some novel physical effects to indicate that the end of the cycle is near. First, many fundamental physical constants of nature, like the strengths of gravity and the strong, weak, and electromagnetic forces, will begin to change noticeably because their values depend on the separation between the branes. They don't change during earlier stages, like today, because the brane separation is frozen. That is why they are interpreted as constants of nature. However, once the branes start to rush toward each other, all the physical constants will start to change in concert. A number of sensitive experiments exist today that monitor these constants and search for time variation. So far, no conclusive evidence for change has been found. According to the cyclic model, physicists performing those same experiments during the last 10 billion years before the next brane collision would detect a large variation of the constants, whose rate would increase as the branes speed up. In the final moments before the crunch, the rapid changes would become dramatic: particles would lose their mass and the laws of nature would be restored to a much simpler and more symmetrical form.

In reality, what is happening is that something enormous is approaching fast along a dimension we cannot see. The realization will come in a flash when, suddenly, everywhere in space lights up with new matter and radiation from the collision. The temperature soars to 10^{15} times the surface temperature of the Sun, evaporating any remnant structures from the previous cycle. The quarks and gluons of which we all are made join the flood of new quarks and gluons created at the bang, and the cycle of the cosmos is renewed.

One Hundred Years

The cosmological debate between the inflationary and the cyclic models is only just beginning to simmer in the scientific community. Many cosmologists have not yet given the issue much consideration because they see no reason for thinking about an alternative until some observation or experiment contradicts the inflationary picture. Others are reluctant to consider a model so deeply rooted in concepts like extra dimensions or branes because they regard these ideas as too far-fetched, even though string theorists are finding these concepts to be essential for unifying our understanding of the fundamental forces. In fact, contemporary versions of the inflationary model are now using the same stringy building blocks.

The reluctance of some to introduce so many new elements into cosmology is understandable. Science usually advances through small variations on an established idea. Radically new directions are not considered unless the scientific case is compelling. For that to happen, the problems with the conventional picture (which might be swept under the rug if there were no competing idea) have to become recognized, and the novel components underlying the new approach have to become familiar. Historically, this conservative approach has served science well, enabling it to make steady progress without getting diverted. In cosmology, for example, the main elements of the current inflationary model—the big bang picture, inflationary expansion, dark matter, and dark energy—were all subject to the same resistance when they were first introduced, and it took many years for them to be accepted.

The cyclic model, if it is worthy, will require similar patience. As discussions and investigations of the cyclic picture continue over the next few years and some of the weaknesses of the inflationary model become more exposed, interest will grow. The fact that two such dissimilar models can predict such similar results is too intriguing to ignore. Creative experimentalists will feel compelled to mount the decisive test between the two views of cosmic history because the issues at stake are too captivating to be ignored.

As the reader now knows, in settling the debate, cosmologists will have come to grips with the most fundamental questions about space and our place in the cosmos; about time and our moment in cosmic history; and about nature and our ultimate ability to figure out its laws. The answers will be our legacy to future generations. One hundred years from now, they will be taught to every schoolchild. They will permeate human discourse and inform our philosophical and religious views. And they will motivate many of the scientific advances of the twenty-second century.

Every elementary science textbook will include the WMAP snapshot or some improved image of the cosmic background radiation across the sky. The authors of the textbook will point to it as one of the great achievements of the twenty-first century. What will they claim about its significance?

If the inflationary model is proven correct, they will write that the image shows the primordial wrinkles created at the end of inflation, about 10^{-35} seconds after the big bang, when the temperature of the universe was about 10^{27} degrees. The universe had a beginning of some sort, perhaps the big bang, but the period of rapid expansion diluted all information about what happened before inflation. Because human-made particle accelerators cannot possibly reach the energies needed to probe conditions before inflation, there is a limit to how much we can learn through observations or experiments about the fundamental laws of the universe. If the inflationary landscape picture survives, it may be impossible to discover the secrets of the universe because everything we see, no matter how far we look, has little in common with the rest of the cosmos, which consists of a combination of inflating regions and pocket universes with different physical properties. As for our own island, the likely outcome is that we are approaching a vacuous, uninhabitable state that will last forever. Perhaps we live in a misanthropic universe.

If the cyclic model proves to be correct, the textbook authors will write that the image shows the splatter of matter and radiation created at the big bang itself. The big bang was not the beginning but the moment separating our current period of expansion and cooling from a previous one. They will explain that the universe has an extra dimension, that the extra dimension is bounded by branes, and that the branes collided with each other to create the bang. They will show how the image can be used to determine the collision speed of the branes and to check that all the matter and radiation we see was created by the collision.

They will write that the WMAP image is also a window on the previous cycle. The small wrinkles in the distribution of matter and energy were created billions of years before by random quantum waves that spontaneously appeared on the surfaces of the branes. A similar effect is beginning now that will eventually give birth to new galaxies and new stars in the next cycle. Because conditions everywhere in the universe are similar to what we observe here and because we can collect observable and measurable traces from an entire cycle, the whole cosmos can be comprehended from our single vantage point.

In 2006, it is too early to say which, if any, of these models will appear in the textbooks of the next century. But all of us can watch as a new theory blossoms into maturity and a mature theory is reinvigorated by the challenge. We can have the fun of debating the two visions of the universe and weighing in with our personal convictions while the matter remains in doubt. And we can do all this secure in the knowledge that the debate will not be endless.

accelerated expansion: an increase in the speed with which space stretches—for example, during inflation and periods when dark energy dominates the universe.

adiabaticity: the condition that the composition of the hot plasma in the early universe is identical everywhere, so that the matter, dark matter, and radiation densities vary across space in the same way; a prediction of both inflationary and cyclic models.

anomaly: a quantum effect that violates a fundamental symmetry, often leading to predictions that are mathematically or physically inconsistent.

anthropic principle: the tenet that the laws and initial conditions that determined the properties of the universe must be consistent with the existence of intelligent life. In recent cosmology discussions, the term refers to a controversial idea of using this tenet as a selection rule for explaining why, in a multiplicity of universes with different physical properties, we ended up in the universe with the particular properties we observe.

antimatter: a substance made of antiparticles. A challenge for cosmology is to explain why there is more matter than antimatter in the universe.

antiparticle: a subatomic constituent of antimatter with the same mass but opposite charge of a matter particle, where combining the antiparticle and particle results in their completely annihilating one another and producing a burst of radiation. For example, the antiparticle of the proton is the antiproton, and the antiparticle of the electron is the positron.

atom: the smallest constituent of a chemical element, composed of protons, neutrons, and electrons.

atomic nucleus: the central object in an atom, made of protons and neutrons.

big bang: the instant when hot matter and radiation were created and space began to expand; in the cyclic model, the instant that the branes collide.

big bang model (standard): the hypothesis that the observable universe emerged from a tiny region of space with nearly infinite density and temperature and has been expanding and cooling ever since. "Standard" usually refers to the original form of the hypothesis, which did not include a period of inflation.

big crunch: in the big bang model, a period of contraction in which space collapses on itself; in the cyclic model, the end of a cycle of evolution in which two branes approach each other and collide.

billion: a thousand million, or 10^9; about 14 billion years have passed since the (most recent) big bang.

brane: derived from "membrane," a basic object in string theory consisting of a one-, two-, or higher-dimensional surface that can move through space, stretch, curve, wiggle, and collide with other similar constituents. In this terminology, a string is a 1-brane, a membrane is a 2-brane, and the three-dimensional space we live in is a 3-brane.

braneworld: a possibility arising from M theory in which the usual three dimensions (height, width, and length) lie within a 3-brane. In the cyclic model, our braneworld collides at regular intervals with a second braneworld separated from it by a finite gap.

bubble (and bubble nucleation): a quantum process in which a field trapped in a false vacuum by an energy barrier spontaneously fluctuates and escapes to the other side of the barrier, producing a spherical volume of true vacuum that subsequently grows and converts the surrounding false vacuum to true vacuum.

chaotic inflation: the notion that the universe emerged from the big bang with a randomly varying distribution of energy, including some rare regions with sufficiently large vacuum energy to undergo an extended period of rapidly accelerating expansion.

conservation of energy: the principle that the total energy of an isolated system does not change over time.

cosmic background radiation (a.k.a. cosmic microwave background): light that has been streaming through the universe since it was first emitted 380,000 years after the big bang, as the first atoms formed in the universe. Today, the radiation has a temperature of 2.73 degrees Kelvin and consists primarily of microwaves.

cosmic strings: hypothetical strandlike concentrations of energy produced by quantum fields, predicted by some grand unified theories.

cosmological constant: the energy of the vacuum, a form of dark energy that is constant in time and perfectly uniformly spread across space.

cosmology: the study of the evolution and composition of the universe.

curvature: a measure of the bending or curving of space or space-time.

cyclic: regularly repeating; in this book, referring to the regular collisions between branes in the cyclic model.

dark energy: the majority of the energy of the universe today, consisting of a component that repels itself gravitationally and causes the expansion of the universe to accelerate.

dark matter: the majority of the mass of the universe, consisting of particles that gravitationally attract one another, just like ordinary matter, but that do not scatter or absorb light.

density: the ratio of the total energy (or mass) to the total volume of a system or part of a system.

density perturbation (or fluctuations): a small spatial variation of the density in the early universe, which, after billions of years, can grow into galaxies and other large-scale structures.

dimensions: the number of coordinates needed to specify the position of a point. For example, height, length, and width are the three dimensions experienced in everyday life. (In some contexts, time is also counted as a dimension.)

Doppler effect: the phenomenon whereby electromagnetic or sound waves from a source moving away from the observer are received at a lower frequency than they were emitted with; conversely, if the source is approaching the observer, the waves are received at a higher frequency.

ekpyrosis: a collision between two branes that produces a flat, expanding universe filled with matter and radiation, with a nearly scale-invariant distribution of density inhomogeneities.

electric field: the field responsible for the attraction or repulsion of electrically charged objects.

electromagnetic waves: waves of oscillating electric and magnetic fields that travel through empty space at 3×10^8 meters per second.

electromagnetism, electromagnetic force: the force that acts between charged particles and that causes the generation and absorption of electromagnetic waves; one of the four fundamental forces of nature (along with the gravitational, strong, and weak forces).

electron: one of the basic constituents of matter, two thousand times lighter than a proton and with an electric charge opposite to that of a proton.

electroweak theory: the theory unifying the electromagnetic force with the weak nuclear force, into the electroweak force.

energy barrier: a hill on the energy curve representing an obstacle to the motion of a field toward the minimum, causing it to get trapped in a state of higher energy.

energy curve: an illustration showing how the energy stored in a field (such as the Higgs field) depends on its value, as shown on page 83.

entropy: the disorder or randomness of a system.

eternal inflation: the notion that, once started, inflation never ends because quantum fluctuations ensure that there are always some regions of space in which the vacuum energy remains high enough for inflation to continue.

extra dimensions: spatial dimensions beyond the usual three (height, width, and depth) experienced in everyday life.

false vacuum: a state empty of all particles and radiation in which at least one field (such as the Higgs field) is trapped by an energy barrier in a state of high energy (e.g., as indicated on an energy curve; see page 89).

field: a quantity having a particular value at each point in space. The value might be a number (as for the inflaton field), a direction (as for an electric or magnetic field), or many numbers (as for a Higgs field breaking a symmetry).

fine-tuning: a term used to describe the unnatural adjustment of constants and initial conditions required for a theory to agree with experiments. Fine-tuning is usually a sign that a theory needs to be revised or replaced.

flat: a term used to describe a universe in which the curvature of space is zero.

flatness problem: the challenge of explaining why the universe is not curved even though the big bang model suggests that this possibility is likely.

force fields: the entities that transmit the gravitational, strong, electromagnetic, and weak forces between matter particles; examples are gravitons (gravity), gluons (strong), photons (electromagnetic), and W and Z bosons (weak).

Friedmann's equation: the equation, derived by Alexander Friedmann from general relativity, that describes how the expansion of the universe depends on the amounts and types of energy in the universe and the curvature of space.

galaxy cluster: a collection of galaxies, ranging in size from small clusters like our Local Group, made of thirty or so galaxies, up to giant galaxy clusters like Virgo, which contains thousands of galaxies.

gaussianity: the condition that the matter and radiation densities vary from place to place according to a Gaussian (or normal) distribution, the bell curve commonly used in statistics; a prediction of both inflationary and cyclic models.

general relativity: Einstein's theory of gravity that predicts that space can curve, stretch, contract, and wiggle.

googol: the number 10^{100} (1 followed by 100 zeroes), roughly the least amount of expansion the universe has to undergo during an inflationary epoch to explain the smoothness and flatness of the universe.

grand unification: the notion that the strong, electromagnetic, and weak forces, which have different strengths and characteristics at low temperatures, emerged from a single symmetrical force as the temperature of the expanding universe fell below 10^{27} degrees Kelvin; the separation into three forces was caused by Higgs fields, whose strength turned on as the universe cooled.

gravitational lensing: the bending of light, which allows us to visualize the distribution of matter, including even dark matter, which we cannot see directly.

gravitational wave: a ripple in space that propagates at the speed of light, causing space to squeeze in one direction and stretch in another as the wave passes.

graviton: the smallest unit of a gravitational wave possible according to the laws of quantum physics; analogous to the photon in electromagnetism.

gravity: one of the four fundamental forces in nature that causes masses to attract one another and space to curve, stretch, and wiggle.

heat: the energy associated with the random motion of particles in a gas, liquid, or solid.

heterotic M theory: the most realistic version of M theory, it posits two branes separated by a gap, and reduces to heterotic string theory as the gap becomes small. M theory motivated the invention of the ekpyrotic and cyclic models.

heterotic string theory: one of the most promising versions of string theory, with enough structure to describe all the forces and particles of nature; related to Hořava-Witten theory and heterotic M theory.

Higgs field: a field that permeates the universe whose value determines masses of elementary particles, breaks the symmetry among the fundamental forces, and fixes the energy of the vacuum.

Higgs strength: the magnitude or absolute value of the Higgs field; the quantity that determines the amount of energy stored in the Higgs field.

Hořava-Witten theory: see *heterotic M theory*.

horizon (or horizon distance): according to the standard big bang picture, the maximum distance that light can have traveled since the big bang or, equivalently, the maximum distance that one can observe.

horizon problem: the challenge of explaining why the distribution of matter and radiation is so uniform throughout the observable universe.

inflation: a period of extraordinarily rapid accelerated expansion that, according to the inflationary model of the universe, is supposed to have smoothed and flattened the universe and created small inhomogeneities that evolved into galaxies.

Kaluza-Klein: the concept, introduced by Theodor Kaluza and Oskar Klein, that there can be extra dimensions of space that are undetectable because they form closed loops that curve back on themselves within a microscopic distance, a key component in string theory.

kinetic energy: energy associated with motion; i.e., the motional energy of a rolling ball, moving brane, or changing Higgs field.

landscape: the notion that string theory may predict an exponentially large, perhaps infinite, number of distinct false vacua with different physical properties and that the observable universe corresponds to one of these possibilities. The concept is invoked to support the idea of a *multiverse* (see page 268) in which the properties of our universe are determined by the *anthropic principle* (see page 259).

magnetic field: the field responsible for the attraction or repulsion of two magnets, or of two wires carrying electric currents.

magnetic monopole: a hypothetical pointlike particle with only one pole of magnetic field (north or south) that is predicted to exist according to grand unified theories and string theory.

matter-dominated epoch: the period of cosmic history when ordinary matter and dark matter made up most of the energy in the universe.

matter particles: the elementary pointlike constituents of matter, such as quarks and electrons.

microwave: an invisible form of electromagnetic radiation with wavelengths ranging from one millimeter to one meter; the type of light making up the cosmic background radiation.

monopole problem: the challenge of explaining why the universe does not contain the high abundance of magnetic monopoles predicted by the grand unified theories and the standard big bang model.

multiverse: the hypothetical notion of a multiplicity of different universes or separate, noninteracting parts of the universe with different physical properties.

neutrino: a nearly massless, electrically neutral particle produced in the early universe and the interior of stars that interacts very weakly with ordinary matter.

neutron: an electrically uncharged particle consisting of an "up" and two "down" quarks, found in atomic nuclei.

nucleus: the tiny, extremely dense object found at the center of atoms, composed of protons and neutrons.

observable universe: the region of the universe close enough to the Earth that light emitted from anywhere within it has had time to reach us.

ordinary matter: matter consisting of protons, neutrons, and electrons.

oscillatory universe: the term often used to refer to earlier attempts at cyclic models in which the usual three spatial dimensions undergo regularly repeating expansion and contraction. (Distinct from the cyclic model discussed in this book, in which only an extra dimension expands and contracts and the usual three dimensions expand from one cycle to the next.)

p-brane: an object with p space dimensions that spans a region of space; for example, a string is a 1-brane and a membrane is a 2-brane.

photon: the smallest unit of electromagnetic radiation (or light) according to quantum physics.

plasma: a substance that filled the early universe, consisting of a hot gas of charged particles: atomic nuclei, electrons, and radiation.

pocket universe: a region of space that has completed a finite period of rapid inflation, filled with matter and radiation, and is now expanding at a much slower rate. According to the inflationary model, we live in a pocket universe and there may be infinitely many others.

polarimeter: an instrument for measuring the polarization of electromagnetic waves, such as light or microwaves.

polarization: in an electromagnetic wave or light, the direction along which the electric field oscillates as the wave propagates through space, as illustrated on page 76.

potential energy: energy stored up in a force or a field.

proton: an electrically charged particle consisting of one "down" and two "up" quarks, found in atomic nuclei.

quantum fluctuations: according to the Heisenberg Uncertainty Principle, which underlies quantum physics, the ever-present random jumps and wiggles of all particles and fields.

quarks: elementary constituents of matter that bind together to form protons and neutrons.

quintessence: a form of dark energy whose density decreases slowly with time and whose distribution in space is not perfectly uniform; the alternative to a cosmological constant for explaining why the expansion of the universe is currently speeding up.

radiation: a form of energy consisting of electromagnetic waves (or light) and massive particles moving at or close to the speed of light.

radiation-dominated epoch: the period in the history of the universe during which the energy per unit volume of radiation was greater than that of all other forms of energy.

recession speed: the speed of an object moving away from an observer.

red shift: the phenomenon whereby the wavelength of light (or other forms of radiation) increases as it travels across the universe.

scale-invariant spectrum: a term describing a special spatial distribution of energy (or matter) that can be expressed as a sum of sinusoidal waves of increasing and decreasing energy (or matter) with different frequencies and with different orientations in three dimensions, but with the same amplitudes (differences between peaks and troughs). Both the cyclic and the inflationary models predict a nearly scale-invariant spectrum of energy and matter in the early universe.

second law of thermodynamics: the principle that the entropy (or disorder) of an isolated system, such as the universe, never decreases. Typically, entropy increases until a system reaches uniform thermal equilibrium.

singularity: a location in space and/or time at which a mathematical description of the physical laws ceases to be valid because certain quantities become infinite.

space-time: the unified picture of space and time that emerged from Einstein's theories of special and general relativity.

special relativity: Einstein's theory of space and time, founded on the principle that the speed of light is the same for all observers. According to the theory, measurements of space and time differ for observers who are moving with respect to one another.

spectrum: the distribution of energy carried by radiation of different wavelengths.

spin: a quantum property of elementary quanta of matter or force fields, such as electrons or photons, describing how they behave like tiny spinning balls, but with a spin that comes only in discrete, quantized units.

standard model: the collection of forces and particles successfully describing all the phenomena seen in particle physics laboratories.

string (a.k.a. superstring): according to string theory, strings are fundamental constituents of nature consisting of one-dimensional vibrating strands. Thus particles perceived as being elementary (like quarks) are actually vibrating strings, with different types of quarks corresponding to different vibrational patterns.

strong nuclear force: the force that holds quarks together in protons and neutrons, and holds protons and neutrons together in atomic nuclei.

supersymmetry (a.k.a. SUSY): a proposed extension of Einstein's theory of special relativity with added relations between space and time that imply a symmetry between matter particles and force fields. This essential mathematical element of string theory cancels some of the infinities commonly encountered in quantum field theory and enhances the simplicity and unity of a theory by relating matter particles and force fields.

symmetry breaking: the process whereby a Higgs field turns on throughout space and breaks the symmetry between the particles and forces.

thermodynamics: the study of how the physical properties of a system depend on the temperature, pressure, and volume.

tilt: a systematic deviation from an exactly scale-invariant noise spectrum of density perturbations; for example, a red tilt means that waves of density with smaller wavelengths have smaller amplitudes, a prediction of inflationary and cyclic models. (The tilt is called "red" because a spectrum of visible light in which waves with smaller wavelengths have less intensity appears red to the eye.)

trillion: a million million (or 1,000,000,000,000); less than the national debt of the United States in dollars and, for the purposes of this book, roughly the number of years between big bangs and, hence, the duration of one cycle.

true vacuum: a state empty of all particles and radiation in which fields (such as the Higgs field) have values corresponding to the lowest possible energy (e.g., as indicated on an energy curve; see page 83).

tunneling: a rare process allowed by quantum physics in which a particle or field (e.g., the Higgs field) trapped by an energy barrier can occasionally fluctuate and escape to the other side of the barrier. In the case where the field is initially trapped in a false vacuum, tunneling occurs through the spontaneous production of bubbles of true vacuum that grow and convert false vacuum to true vacuum.

unified theory: an attempt to describe many disparate phenomena, such as particles and forces, within a single, simpler framework.

vacuum energy: the energy of empty space, in which there is no matter or radiation.

wavelength: the distance between successive crests (or troughs) of a wave.

WMAP: an acronym for the Wilkinson Microwave Anisotropy Probe, a satellite orbiting a million miles from the Earth that gathers light emitted 14 billion years ago (the cosmic background radiation) to form an image of the early universe.

Below is a list of works that may be useful for readers who want to further explore various aspects of the science addressed in this book.

Barrow, John, and Frank J. Tipler. *The Anthropic Cosmological Principle.* Oxford: Oxford University Press, 1988.

Bernstein, Jeremy, and Gerald Feinberg. *Cosmological Constants: Papers in Modern Cosmology.* New York: Columbia University Press, 1986.

Bondi, Hermann. *Relativity and Common Sense.* London: Heinemann, 1965.

Bostrom, Nick. *Anthropic Bias.* London: Routledge, 2003.

Cole, K. C. *The Hole in the Universe.* New York: Harcourt, 2001.

Davies, Paul. *About Time.* New York: Simon & Schuster, 1995.

————. *The Last Three Minutes.* New York: Basic Books, 1994.

————. *The New Physics.* Cambridge, U.K.: Cambridge University Press, 1992.

Deutsch, David. *The Fabric of Reality.* New York: Allen Lane, 1997.

Ferreira, Pedro. *The State of the Universe.* London: Cassell, 2006.

Ferris, Timothy. *The Whole Shebang.* New York: Simon & Schuster, 1997.

Gell-Mann, Murray. *The Quark and the Jaguar.* New York: W. H. Freeman, 1994.

Gott, J. Richard. *Time Travel in Einstein's Universe.* Boston: Houghton Mifflin, 2001.

Greene, Brian. *Elegant Universe.* New York: Vintage, 2000.

————. *Fabric of the Cosmos.* New York: Vintage, 2005.

Gribbin, John. *The Birth of Time.* New Haven: Yale University Press, 2001.

————. *In the Beginning*. New York: Little, Brown, 2005.

Guth, Alan. *The Inflationary Universe*. Reading, Mass.: Perseus, 1997.

Hawking, Stephen. *A Brief History of Time*. New York: Bantam, 1988.

————. *The Universe in a Nutshell*. New York: Bantam, 2001.

Hawking, Stephen, and Roger Penrose. *The Nature of Space and Time*. Princeton: Princeton University Press, 1996.

Kaku, Michio. *Hyperspace*. New York: Oxford University Press, 1994.

————. *Parallel Worlds*. New York: Doubleday, 2004.

Kirshner, Robert. *The Extravagant Universe*. Princeton: Princeton University Press, 2002.

Kragh, Helge. *Cosmology and Controversy*. Princeton: Princeton University Press, 1996.

Krauss, Lawrence. *Hiding in the Mirror*. New York: Penguin, 2005.

————. *Quintessence*. New York: Perseus, 2000.

Lederman, Leon M., and David N. Schramm. *From Quarks to the Cosmos*. New York: W. H. Freeman, 1989.

Levin, Janna. *How the Universe Got Its Spots*. London: Weidenfeld & Nicolson, 2002.

Mahon, Basil. *The Man Who Changed Everything: The Life of James Clerk Maxwell*. London: John Wiley, 2003.

Overbye, Dennis. *Lonely Hearts of the Cosmos*. New York: HarperCollins, 1991.

Pais, Abraham. *Subtle Is the Lord*. Oxford: Oxford University Press, 1982.

Penrose, Roger. *The Road to Reality*. New York: Knopf, 2005.

Randall, Lisa. *Warped Passages*. New York: HarperCollins, 2005.

Rees, Martin. *Before the Beginning*. Reading, Mass.: Addison-Wesley, 1997.

————. *Just Six Numbers*. New York: Basic Books, 2001.

Russell, Bertrand. *The ABC of Relativity*. London: Allen & Unwin, 1985.

Silk, Joseph. *The Infinite Cosmos*. Oxford: Oxford University Press, 2006.

Singh, Simon. *Big Bang*. New York: HarperCollins, 2004.

Smolin, Lee. *Three Roads to Quantum Gravity*. New York: Basic Books, 2001.

Susskind, Leonard. *The Cosmic Landscape*. New York: Little, Brown, 2005.

Thorne, Kip. *Black Holes and Time Warps*. New York: W. W. Norton, 1994.

Tyson, Neil deGrasse, and Donald Goldsmith. *Origins*. New York: W. W. Norton, 2004.

Vilenkin, Alex. *Many Worlds in One*. New York: Hill & Wang, 2006.

Weinberg, Steven. *Dreams of a Final Theory*. New York: Pantheon, 1992.

————. *The First Three Minutes*. New York: Basic Books, 1993.

Wilczek, Frank. *Fantastic Realities*. Singapore: World Scientific, 2006.

Zee, Anthony. *Fearful Symmetry*. Princeton: Princeton University Press, 1999.

PAUL J. STEINHARDT is the Albert Einstein Professor in Science at Princeton University, on the faculties of both the physics and astrophysics departments. He is one of the leading architects of the inflationary and cyclic theories of the universe, the two competing views of the origin, evolution, and future of the universe that are the focus of *Endless Universe*. He is also recognized for contributions in condensed matter physics, particularly the discovery of quasicrystals. In addition to his research, he has lectured at both a general and technical level throughout the world and written numerous popular articles. Steinhardt and his wife, Nancy, have four children and live in Princeton, New Jersey.

NEIL TUROK holds the Chair of Mathematical Physics (1967) in the Department of Applied Mathematics and Theoretical Physics at Cambridge University. He is a world leader in the development and testing of fundamental physical theories of the cosmos, such as those described in *Endless Universe*, and he is coauthor of the new cyclic theory. His research encompasses cosmology, particle physics, mathematical physics, astrophysics, and string theory. A native of South Africa, Turok recently founded the African Institute for Mathematical Sciences (AIMS) in Cape Town, a postgraduate center supporting the development of mathematics and science across the African continent. Turok lives in Cambridge, UK, with his partner, Corinne, his daughter, Ruby, and a variable number of other animals.